Digital Circuits: Volume 3
Flip-Flops, Counters, Shift Registers, Decoders, Mutiplexers
Engineer's Tutor Series

by

Amalou   Abdelilah

Weber Systems, Incorporated
Chesterland, Ohio

Published by:
Weber Systems, Inc.
8437 Mayfield Road
Chesterland, Ohio 44026
(216) 729-2858

For information on translations and book distributors outside of the United States, please contact WSI at the above address.

Digital Circuits: Volume 3: Flip-Flops, Counters, Shift Registers, Decoders, Multiplexers

Manufactured in the U.S.A
ISBN: 0-929704-05-3

Contents

CHAPTER 3

SHIFT REGISTERS 109

CHAPTER 4

DECODERS/ENCODERS 145

CHAPTER 5

MULTIPLEXERS/DEMULTIPLEXERS 189

Preface

The Engineer's Tutor Series is designed for the engineering student who is experiencing difficulty with his coursework or the engineer or technician who wishes to master a subject on his own. Digital Circuits Volume 3 is a part of a 3 volume set which also includes:

- Digital Circuits Volume 1: Numbering Systems, Binary Codes, Logic Gates, Boolean Algebra

- Digital Circuits Volume 2: Truth Tables, Minterms, Maxterms, Karnaugh Maps

Volume 3 consists of an expanded discussion of the intermediate subjects covered in a digital circuits course: flip-flops, counters, shift registers, decoders, and multiplexers. The depth of the discussion is much greater than that found in typical digital circuits textbooks. This expanded discussion will assist the student who is experiencing difficulty in mastering these topics.

This book's subject matter is presented in an easily understood format. Topics are first discussed in a general sense followed by one or more examples which illustrate that topic in practical terms. For example, in the section on flip-flops, R-S flip-flops are first discussed in a general sense. Four examples then follow which illustrate R-S flip-flop operation. A step-by-step explanation is provided with each example so the reader fully understands R-S flip-flop operation.

With its detailed explanations and numerous examples and solved problems, this book is ideal for the student who is having difficulty mastering his digital circuits coursework or the engineer or technician who wishes to learn this process through independent study.

About This Book
This book is divided into five separate chapters. Each chapter includes numerous examples and solved problems. A short description of each chapter follows.

1. Flip-Flops
Chapter 1 consists of a detailed discussion of flip-flops including R-S, D, J-K, T, and master-slave combinations. Each flip-flop's operation is described in a step-by-step manner. Transition tables, pulse trains, and examples are included in each individual discussion. Edge-triggering and timing are also included in this chapter. Several experiments are included which demonstrate flip-flop operation.

2. Counters
The operation and design of binary BCD and parallel counters are described in chapter 2. Each counter is illustrated and its operation is described step-by-step. Numerous examples and solved problems are included to illustrate the concepts being discussed. An experiment is included where the student builds and operates a mod-6 up counter.

3. Shift Registers
Shift registers are discussed in chapter 3. Serial in-parallel out and parallel shift registers are described as are arithmetic operations using shift registers. The chapter concludes with an experiment where the student builds and operates a 4-bit right shift register.

4. Decoders/Encoders
Chapter 4 includes a discussion of decoders and encoders, including BCD to decimal, BCD to 7 segment, and the 74147 encoder. An experiment is included where the reader operates and tests the 74ALS139 decoder.

5. Multiplexers/Demultiplexers
Multiplexers and demultiplexers are described in chapter 5. A step-by-step description of multiplexer operation is included along with examples and solved problems which further illustrate multiplexer operation. Practical applications of common multiplexers such as parallel/serial conversion, binary word generation, and Boolean function generation are also discussed.

1

Flip-Flops

Introduction

Digital circuit are categorized as combinational logic circuits and sequential logic circuits.

To this point we have designed combinational logic circuits. Combinational logic circuits are circuits that have no memory capacity. In other words they do not "remember" past signals (or inputs). A combinational logic circuit's operation depends only on the present inputs and won't be affected by past inputs.

A sequential logic circuit can remember past inputs by feeding back its output. The process of taking an output signal and using it as an input to the same circuit creates a memory in the circuit as we'll see later. A sequential circuit's memory will affect and control its subsequent behavior. Sequential logic circuits are called 'sequential' circuits because signals (or inputs) should be applied in a certain 'sequence' to obtain a specific or desired behavior.

Sequential logic circuits are divided into two groups: *synchronous* and *asynchronous*.

In a synchronous sequential logic circuit, all the components (or ICs) respond to input signals simultaneously, because they are synchronized by a master clock. A *master clock* is a device that generates pulses to the synchronous sequential logic circuit. The components of the circuit respond to their inputs only when allowed by the master clock; otherwise the components are disabled and ignore their input signals.

For a synchronous sequential logic circuit to operate properly, the master clock frequency should be set to the response time (or delay) of the slowest component in the circuit, so that all the circuit's components can respond to their inputs. Keep in mind that setting the master clock to the response frequency of the slowest component slows the overall operation of the logic circuit. (We'll discuss this concept in greater detail in subsequent sections.)

In an asynchronous sequential logic circuit, each component responds at its own frequency. The overall operation of the circuit is faster. The design of asynchronous sequential logic circuit is very difficult and in many cases asynchronous design is impractical.

Most digital computers use synchronous sequential logic circuits. Nevertheless asynchronous logic circuits or a combination of synchronous and asynchronous sequential circuits are found in digital computers.

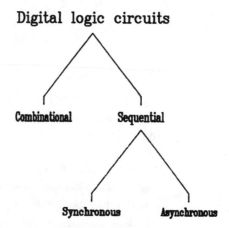

Flip-Flops
Definition

A *flip-flop* is a digital circuit made up of a number of logic gates (*see* figure 1.1). A flip-flop is also referred to as a bi-stable multivibrator or a latch. *The main function of a flip-flop is memory.* A flip-flop is used to store a binary bit (0 or 1). A flip-flop's output is called its *state*. If the state (or output) of the flip-flop is binary 0, then we say that the flip-flop is storing a binary bit 0. When the state of the flip-flop is binary 1, then the flip-flop is storing a binary bit 1.

A flip-flop's output is usually referred to as Q. Since Q represents the state of the flip-flop, Q represents the binary bit stored in the flip-flop. (*See* figure 1.1.)

$\overline{Q}$ represents the complement of Q which is the flip-flop's other possible output. Thus if $\overline{Q} = 0$ then Q = 1 and vice versa. (*See* figure 1.1.)

Flip-flops can assume one or more inputs. The name assigned to the input will vary depending on the type of flip-flop. (*See* figure 1.2.)

The main function of a flip-flop is to store a binary bit (0 or 1). Once a binary bit is stored, the flip-flop remains unchanged until an appropriate signal is applied to its input(s). (We'll discuss this concept in more detail later in this chapter.)

There are many types of flip-flops. The most common ones are: R-S flip-flop, D flip-flop, J-K flip-flop, and T flip-flop (*see* figure 1.2.). The operation of these flip-flops will be detailed in the following sections.

In later chapters of this book, we'll learn how flip-flops can be grouped together to form counters, registers, ..etc.

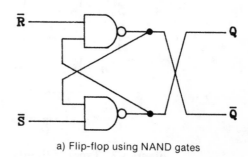

a) Flip-flop using NAND gates

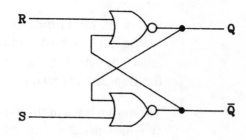

b) Flip-flop using NOR gates

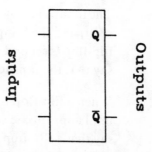

c) Logic symbol of a flip-flop

Figure 1.1. Flip-flops

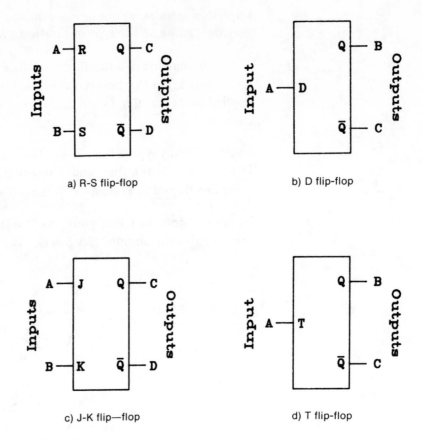

a) R-S flip-flop   b) D flip-flop

c) J-K flip—flop   d) T flip-flop

Figure 1.2. Types of flip-flops

R-S Flip-flop

One of the most fundamental flip-flops is the R-S flip-flop. The R-S flip-flop can be implemented using either NAND or NOR gates (*see* figure 1.1). The most popular R-S flip-flops utilize NOR gates. We'll discuss NOR gated flip-flops in the remaining chapters of this book.

The output leads of the R-S flip-flop are called Q and $\overline{Q}$. $\overline{Q}$ is the complement. Q represents the state of the flip-flop. By examining output Q we can determine what binary bit is stored by the flip-flop. If Q=1, the flip-flop is storing binary bit 1 and if Q=0, the flip-flop is storing binary bit 0.

The input leads of the R-S flip-flop are called R and S. The R input stands for reset, and the S input stands for set. When the output Q is 1, we say that the flip-flop is set and that it is storing a binary bit 1. When the output Q is 0, we say that the flip-flop is reset and that it is storing a binary bit 0.

When a flip-flop is in one of its states (set or reset), it will remain unchanged until appropriate signals or pulses are applied to its inputs. Therefore we can say that a flip-flop has a memory characteristic or that a flip-flop remembers past signals (or binary bits). R-S flip-flop operation can be described as follows:

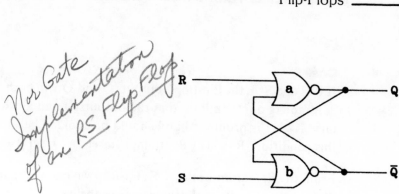

Figure 1.3. R-S flip-flop

Case 1

When Q=0, R=0, and a high pulse level (or logic 1) is applied to the S input, we have the following operation:

Inputs to gate b:	0 1 (Q and S)
Output from gate b:	0 (not Q=0)
Inputs to gate a:	0 0 (not Q and R)
Output from gate a:	1 (Q=1)

The state of the flip-flop will change and Q will become 1. We say that the flip-flop has been set (Q=1).

Case 2

If Q=0, R=0 and a low pulse level (or logic 0) is applied to the S input, the state of the flip flop will remain unchanged (Q=0).

Case 3

If Q=1, R=0 and a high pulse level is applied to the S input, the flip-flop will remain set (Q=1).

Case 4

If Q=1, R=0 and a low pulse level is applied to the S input, the flip-flop will remain set (Q=1).

Case 5

Now if Q=0, S=0 and a high pulse level is applied to the R input, the flip-flop will remain reset (Q=0).

Case 6

If Q=0, S=0 and a low pulse level is applied to the R input, the flip-flop will remain reset (Q=0).

Case 7

If Q=1, S=0 and a high pulse level is applied to the R input, the state of the flip-flop will change. The flip-flop will be reset and the output Q will become 0.

Case 8

If Q=1, S=0 and a low pulse level is applied to the R input, the flip-flop will remain set (Q=1).

Case 9

If R=1 and S=1, the flip-flop's outputs Q and $\overline{Q}$ will both be low or 0. This state contradicts our definition that the outputs Q and $\overline{Q}$ should be complementary. R=S=1 is prohibited and should be avoided because the flip-flop under this condition (R=S=1) will attempt to set and reset itself simultaneously.

From the operation of the R-S flip-flop we can see that this R-S flip-flop is *high true active*. By high true we mean the a high level input will change the state of the flip-flop if applied to the appropriate input. For example imagine that the flip-flop is set (Q=1) and that a high pulse is applied to the R input (with S=0 so that the prohibited state is avoided); the flip-flop will be reset (Q=0). If the flip-flop is reset (Q=0) and a high level pulse is applied to the S input (with R=0), then the flip-flop will be set (Q=1). A low pulse will not change the state of a flip-flop.

When a flip-flop is set (Q=1) and a high level pulse is applied to the S input (with R=0), the state of the flip-flop remains unchanged because the flip-flop was already set.

When a flip-flop is reset (Q=0) and a high level pulse is applied to the input R (with S=0), the flip-flop remains reset because the high input level on the R input reset the flip-flop a second time leaving the output unchanged (Q=0).

To summarize the operation of a sequential circuit, we use a transition table. A transition table is the equivalent of a truth table for sequential circuits. A transition table represents the states of the flip-flop before and after a pulse clock is applied (*see* table 1.1).

Table 1.1. Transition table

Before Pulse		Inputs		After Pulse	
Q(t)		R	S	Q(t+t1)	
0	(reset)	0	0	0	(reset)
0	(reset)	0	1	1	(set)
0	(reset)	1	0	0	(reset)
0	(reset)	1	1	not allowed state	
1	(set)	0	0	1	(set)
1	(set)	0	1	1	(set)
1	(set)	1	0	0	(reset)
1	(set)	1	1	not allowed state	

Note that Q(t) represents the present time or the time before the pulse is applied. Q(t+t1) represents the time after the pulse is applied.

When *R=S=0*, the state of the flip-flop remains the same. This is a useful concept in designing circuits as we'll see later in this book.

Another table called an *excitation table* is used in sequential circuits to show what signal should be applied to the inputs to change (or to keep) the state of a flip-flop (*see* table 1.2)

Table 1.2. Excitation table

State change		Inputs	
From	To	R	S
0	0	X	0
0	1	0	1
1	0	1	0
1	1	0	X

Note that X is a don't care case; X can be represented by either binary 0 or 1. For example, a state change from 0 to 0 would require that the S input be 0, and the R input could be either 0 or 1 (*see* cases 2 and 5). If the state change is from 1 to 1, the R input should be equal to 0, and the S input could be either 0 or 1 (*see* cases 3 and 4).

With AND Gates the flip flop is active low.

Example

With a NOR gate implementation, does an R-S flip have active *low* or *high* inputs? Explain.

Solution

With NOR gate implementation an R-S flip-flop has active high inputs. This means that an appropriate high level input will change the high input state of the flip-flop.

Example

Fill in the blank spaces with an appropriate answer.

1) If Q=0, R=0 and a high level pulse is applied to the input S, the Q output becomes _____(0 or 1), and the flip-flop is said to be_____(set or reset).

2) If Q=1, S=0 and a high level pulse is applied to the input R, the not Q output becomes _____(0 or 1), and the flip-flop is said to be storing binary bit _____(0 or 1).

Solution

1) If Q=0, R=0 and a high level pulse is applied to the input S, the Q output becomes *1*, and the flip-flop is said to be *set*.

2) If Q=1, S=0 and a high level pulse is applied to the input R, the not Q output becomes *1*, and the flip-flop is said to be storing binary bit *0*.

Example

What is the output pulse Q of the following flip-flop? Also, list the $\overline{Q}$ pulse.

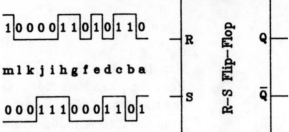

Solution

	Inputs		Outputs		State of flip-flop
Pulse	R	S	Q	$\overline{Q}$	
a	0	1	1	0	Set
b	1	0	0	1	Reset
c	1	1	Not allowed		N/A
d	0	1	1	0	Set
e	1	0	0	1	Reset
f	0	0	0	1	Reset
g	1	0	0	1	Reset
h	1	1	Not allowed		N/A
i	0	1	1	0	Set
j	0	1	1	0	Set
k	0	0	1	0	Set
l	0	0	1	0	Set
m	1	0	0	1	Reset

Note that pulses f, k, and l do not change the state of the flip-flop, proving the preceding state is repeated when R=S=0.

The output pulses are:

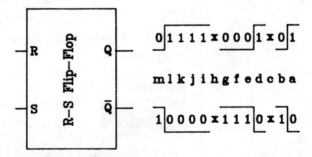

Note that X means "not defined."

Practice Problems

Problem 1.1

Is a *NAND implementation R-S flip-flop* active low or active high? Why?

Solution

A flip-flip implemented with NAND gates is *low active* because of the operation of the NAND gate (*see* Digital Circuits: Volume 1). An R-S flip-flop implemented with a NAND gate is depicted below.

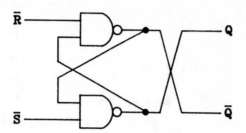

We see that a low level pulse on the R input (with S=1) would reset the flip-flop (Q(t)=0), and a low level pulse on the S input (with R=1) would set the flip-flop (Q(t)=1). When both inputs, R and S, are high (R=S=1), the flip-flop's state doesn't change and Q(t)=Q(t+t1). Finally when R=S=0 the flip-flop is trying to be set and reset simultaneously and Q(t)= $\overline{Q}$(t)=1 which would contradict the fact that Q and $\overline{Q}$ should always be complementary. Therefore the condition of R=S=0 is not allowed.

Problem 1.2

What is the output pulse train of the following NOR *flip-flop* implementation? Use table 1.1.

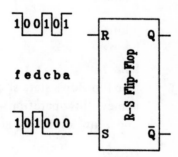

Solution

Pulse a: R=1 and S=0   Q=0 (*see* case 5 or 7)
Pulse b: R=0 and S=0   Q=0 (same as pulse a, *see* case 2)
Pulse c: R=1 and S=0   Q=0 (*see* case 5)
Pulse d: R=0 and S=1   Q=1 (*see* case 1)
Pulse e: R=0 and S=0   Q=1 (same as pulse d, *see* case 4)
Pulse f: R=1 and S=1   Not allowed (*see* case 9)

The output pulse train resembles the following illustration:

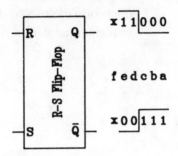

Problem 1.3

Complete the following table (for a NOR implementation flip-flop):

	Q(t)	R	S	Q(t+t1)	State after pulse (set or reset)
Line 1	0	...	...	1	...
Line 2	1	...	...	0	...
Line 3	1	...	...	1	...
Line 4	0	...	...	0	...
Line 5	0	1	...	1	...
Line 6	...	...	1	...	set

Solution

	Q(t)	R	S	Q(t+t1)	State after pulse (set or reset)
Line 1	0	0	1	1	set
Line 2	1	1	0	0	reset
Line 3	1	0	X	1	set
Line 4	0	X	0	0	reset
Line 5	X	0	1	1	set

Line 1
The flip-flop's state should change from reset (Q(t)=0) to set (Q(t+t1)=1). To achieve this operation a high level pulse should be applied to the S input (S=1) and a low level pulse should be applied to the R input (R=0). (*See* case 1.)

Line 2
The flip-flop's state should change from set (Q(t)=1) to reset (Q(t+t1)= 0). To obtain this operation a low level pulse should be applied to the S input (S=0), and a high level pulse should be applied to the R input (R=1). (*See* case 7.)

Line 3
The state of the flip-flop should remain set. Thus we set the R input low (R=0) so that the flip-flop won't reset, but the S input can be low or high (S=X). (*See* cases 3 and 4.)

Line 4

Same as line 3, but the state of the flip-flop should remain reset. We apply a low level pulse to the S input (S=0) so that the flip-flop won't be set. The R input could accept any pulse (low or high); the flip-flop won't be affected.

Line 5

Because a high level pulse is applied to the S input a low level pulse should be applied to the R input so that the condition R=S=1 won't occur (*see* case 9). Using the condition R=0, S=1 would make Q(t+t)=1 which sets the flip-flop (*see* cases 1 and 3). Both Q(t)=1 or Q(t)=0 won't affect the final state of the flip-flop; therefore a don't care state could be used and Q(t)=X. (*See* cases 1 and 3.)

Problem 1.4

Complete the following table (NOR implementation flip-flop):

	$\overline{Q}(t)$	R	S	$\overline{Q}(t+t1)$	State after pulse (set or reset)
Line 1	0	0	1	...	.....
Line 2	0	...	...	1	.....
Line 3	1	1	1	...	.....
Line 4	1	0	0	...	.....
Line 5	..	0	0	...	reset
Line 6	0	...	...	1	.....
Line 7	1	...	...	0	.....
Line 8	1	...	...	1	.....
Line 9	0	...	...	0	.....

Solution

	$\overline{Q}(t)$	R	S	$\overline{Q}(t+t1)$	State after pulse (set or reset)
Line 1	0	0	1	0	set
Line 2	0	1	0	1	reset
Line 3	1	1	1	N/A	N/A
Line 4	1	0	0	1	reset
Line 5	1	0	0	1	reset
Line 6	0	1	0	1	reset
Line 7	1	0	1	0	set
Line 8	1	X	0	1	reset
Line 9	0	0	X	0	set

Line 1

If $\overline{Q}(t)=0$ then Q(t)=1. Now by applying a high pulse to the $\hat{S}$ input (with R=0) we set the flip-flop and Q(t+t1)=1. Therefore $\overline{Q}(t+t1)=0$. (*See* case 3).

Line 2

If $\overline{Q}(t+t1)=1$ then Q(t+t1)=0; therefore the flip-flop is reset. To reset a flip-flop that was set (Q(t)=1), a high pulse should be applied to the R input (with S=0). (*See* case 7).

Line 3

R=S=1 not allowed inputs. (*See* case 9).

Line 4

When R=S=0 no change occurs. In our case the flip-flop was reset before the pulses were applied ($\overline{Q}$(t)=1). Therefore flip-flop remains reset and ($\overline{Q}$ (t+t1)=1. (*See* case 2)

Line 5

Same logic level inputs as line 4. Therefore the flip-flop remains reset and Q(t+t1)=1. (*See* case 2.)

Line 6

$\overline{Q}$(t)=0 thus Q(t)=1. After the pulse $\overline{Q}$(t+t1)=1; thus Q(t+t1)=0. Therefore the flip-flop has gone from set to reset and the inputs should be R=1 and S=0. (*See* case 7).

Line 7

$\overline{Q}$(t)=1; thus Q(t)=0. After the pulse $\overline{Q}$(t+t1)=0; thus Q(t+t1)=1. Therefore the flip-flop has gone from reset to set and the inputs required for such an operation are: R=0 and S=1. (*See* case 1.)

Line 8

$\overline{Q}$(t+t1)=1 thus Q(t+t1)=0 and the flip-flop is reset. Before the application of the pulse, the flip-flop was reset (Q(t)=0). Therefore the state of the flip-flop has not been changed. The required input pulses for such an operation are S=0 (if we use S=1 we will be setting the flip-flop) and R=X (with X equal to 0 or 1). R=X is used because of the following reasons:

1) If R=0, the inputs R and S will be 0 and from cases 2 and 4 we know that when this condition (R=S=0) occurs the flip-flop remains unchanged. In our case the flip-flop was reset and will remain reset after the application of the pulse. (*See* case 4.)

2) If R=1 (with S=0), the flip-flop will be automatically reset, thus its first (before application of pulse) state won't change. (*See* line 6.)

We have seen that both conditions R=0 and R=1 (with S=0) worked. Thus if S=0 the pulse to input R would not matter and an X character is used as a "don't care" input.

Line 9

Here we have the same effect as in line 9 with the flip-flop remaining set. First we have to make the R input low (R=0) so that the flip-flop won't be reset. For the S input it doesn't matter what logic level (low or high) is applied because with a low level pulse applied to the S input, S=R=0, the flip-flop will remain unchanged (set), and with a high level pulse applied to the S input, S=1 and R=0, the flip-flop will be set. In both cases, the flip-flop will be set.

Clocked R-S Flip-Flop

The R-S flip-flop discussed in the previous section responds to input signals only when those input signals are actually sensed. Therefore we say that the flip-flop discussed is asynchronous. In other words the R-S flip-flop is not controlled by a master clock and will respond when it senses an incoming input signal.

A clocked R-S flip-flop is one that only responds when the master clock permits it to. A clocked R-S flip-flop is a synchronous device that can be synchronized by a master clock.

Figure 1.4 shows a clocked R-S flip-flop. Note that when the clock pulse is low, the output from gate a and gate b is 0 no matter what signals are applied to the R and S inputs. Thus the inputs to the R-S flip-flop are 0 (from gate a) and 0 (from gate b). Previously, we learned that when we apply 0 to both inputs of an R-S flip-flop, the state of the flip-flop doesn't change. In other words the flip-flop state stays the same because it ignores its inputs.

Going back to our clocked R-S flip-flop, we can say that when CP=0 the R-S flip-flop ignores its inputs. CP=0 actually disables the flip-flop.

If CP=1, the signal output from gates a and b will be identical to the signals on the R and S inputs respectively. In general we can say that when CP=0, the signal from the R and S inputs do not propagate through gates a and b. Therefore the state of the flip-flop remains the same and we say that the flip-flop is disabled by the clock pulse (CP). When CP=1, the R and S input signals propagate through gates a and b, and the R-S flip-flop's state may change. We say that the R-S flip-flop is enabled by the clock pulse (CP).

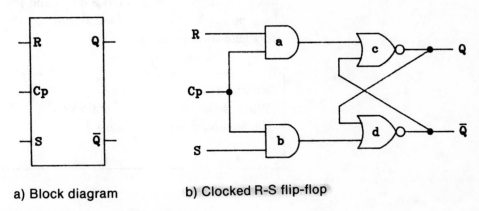

a) Block diagram   b) Clocked R-S flip-flop

Figure 1.4. Clocked R-S flip-flop

We can summarize the operation of a clocked R-S flip-flop in the following table:

Table 1.3. Clocked R-S flip-flop's operation
(NOR gate implementation)

CP	Q(t)	R	S	Q(t+t1)	Final State
0 (disabled)	0	0	0	0	reset
0 (disabled)	0	0	1	0	reset
0 (disabled)	0	1	0	0	reset
0 (disabled)	0	1	1	0	reset
0 (disabled)	1	0	0	1	set
0 (disabled)	1	0	1	1	set
0 (disabled)	1	1	0	1	set
0 (disabled)	1	1	1	1	set
1 (enabled)	0	0	0	0	reset
1 (enabled)	0	0	1	1	set
1 (enabled)	0	1	0	0	reset
1 (enabled)	0	1	1	not allowed	
1 (enabled)	1	0	0	1	set
1 (enabled)	1	0	1	1	set
1 (enabled)	1	1	0	0	reset
1 (enabled)	1	1	1	not allowed	

Q(t): Output of flip-flop before the application of pulse.

Q(t+t1): Output of flip-flop after the application of pulse.

From table 1.3, we see that when CP=0, the flip-flop's state doesn't change. Compare Q(t) and Q(t+t1)). They are identical regardless of what input signals are applied.

When CP=1, the clocked R-S flip-flop operates normally as described previously.

Note that the clock pulse (CP) should be long enough to allow the flip-flop's state to change. Each gate has a delay and for an appropriate operation, the clock pulse should be at least as long as the delay so that the signal can propagate through the gate.

Example

What is the output pulse train of the following flip-flop? (NOR gate implementation). Use table 1.3.

(cont. on page 15)

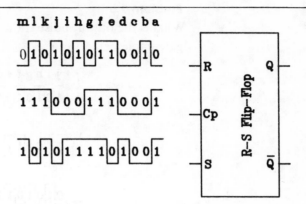

Solution

a: R=0, S=1 and CP=1 (enabled): Q=1, flip-flop is SET
b: R=1, S=0 and CP=0 (disabled): Q=1, flip-flop remains SET
c: R=0, S=0 and CP=0 (disabled): Q=1, flip-flop remains SET
d: R=0, S=1 and CP=0 (disabled): Q=1, flip-flop remains SET
e: R=1, S=0 and CP=1 (enabled): Q=0, flip-flop is RESET
f: R=1, S=1 and CP=1 (enabled): not allowed
g: R=0, S=1 and CP=1 (enabled): Q=1, flip-flop is SET
h: R=1, S=1 and CP=0 (disabled): Q=1, flip-flop remains SET
i: R=0, S=1 and CP=0 (disabled): Q=1, flip-flop remains SET
j: R=1, S=0 and CP=0 (disabled): Q=1, flip-flop remains SET
k: R=0, S=1 and CP=1 (enabled): Q=1, flip-flop is SET
l: R=1, S=0 and CP=1 (enabled): Q=0, flip-flop is RESET
m: R=0, S=1 and CP=1 (enabled): Q=1, flip-flop is SET

Note that when CP=0, the state of the flip-flop is the same as the preceding state and the flip-flop is disabled (pulses b, c, d, h, i, j). When CP=1, the flip-flop is enabled (pulses a, e, f, g, k, l , m).

The output pulse is:

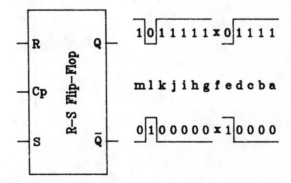

X represents a not allowed state (Q=$\overline{Q}$).

Practice Problems

Problem 1.5

What is the output pulse train of the following flip-flop? (Use table 1.3 NOR gate implementation.)

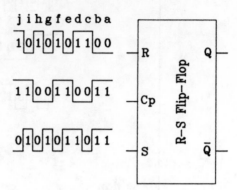

Solution

a: R=0 S=1   CP=1: Q(t+1)=1   SET
b: R=0 S=1   CP=1: Q(t+1)=1   SET
c: R=1 S=0   CP=0: Q(t+1)=1   SET
d: R=1 S=1   CP=0: Q(t+1)=1   SET
e: R=0 S=1   CP=1: Q(t+1)=1   SET
f: R=1 S=0   CP=1: Q(t+1)=0   RESET
g: R=0 S=1   CP=0: Q(t+1)=0   RESET
h: R=1 S=0   CP=0: Q(t+1)=0   RESET
i: R=0 S=1   CP=1: Q(t+1)=1   SET
j: R=1 S=0   CP=1: Q(t+1)=0   RESET

Pulse c should reset the flip-flop but because CP=0, the flip-flop ignores its inputs and remains SET. Pulse d is not allowed but because CP=0, the flip-flop ignores its inputs and therefore remains SET. Pulse g should set the flip-flop but because CP=0, the flip-flop ignores its inputs and remains RESET.

The output pulse train is:

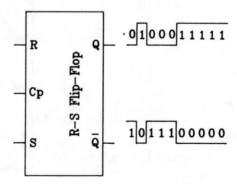

Problem 1.6

Complete the following table for a clocked R-S flip-flop:

	CP	Q(t)	R	S	Q(t+t1)	Comments
Line 1	0	0	1	0	---	----
Line 2	0	---	1	0	1	----
Line 3	---	1	1	0	1	----
Line 4	---	0	0	1	0	----
Line 5	0	1	---	---	----	----
Line 6	0	---	---	---	0	----
Line 7	---	0	1	0	0	----
Line 8	1	---	1	0	---	----

Solution

	CP	Q(t)	R	S	Q(t+t1)	Comments
Line 1	0	0	1	0	0	When CP=0, the flip-flop is disabled and Q(t)=Q(t+t1).
Line 2	0	1	1	0	1	Because CP=0 we know that Q(t)=Q(t+t1), thus Q(t)=1.
Line 3	0	1	1	0	1	We have R=1 and S=0; thus the flip-flop should be reset. Nevertheless the flip-flop is not reset. Since CP=0, it ignores its inputs and remains set.
Line 4	0	0	0	1	0	We have R=0 and S=1. Thus the flip-flop should be set. Nevertheless the flip-flop is not reset. Since CP=0, it ignores its inputs and remains reset.
Line 5	0	1	X	X	1	With CP=0, and Q(t+t1)=Q(t)=1 inputs R and S wouldn't matter.
Line 6	0	0	X	X	0	With CP=0, and Q(t+t1)=Q(t)=0 inputs R and S wouldn't matter.
Line 7	X	0	1	0	0	With Q(t)=Q(t+t1)=0, R=1, and S=0, the flip-flop will reset. Therefore we can't tell if CP=0 or CP=1. Both cases are valid.
Line 8	1	X	1	0	0	Since CP=1, the flip-flop is enabled. R=1 and S=0 reset the flip-flop. Q(t) is of no consequence here.

D Flip-Flop

In the earlier sections we have learned that a clocked R-S flip-flop's output state is undetermined when R=S=1. It would greatly simplify the design process if we could obtain a clocked flip-flop with no undetermined states.

One way of doing this is to prevent situations were R=S=1 by making the inputs to the flip-flop (R and S) always complementary. We can accomplish this by adding an inverter gate between the R and S inputs (*see* figure 1.5). The result is a clocked flip-flop that has one input that controls both R and S inputs. Since R and S are never the same, the undetermined state will never occur. This implementation is called a *D flip-flop*.

A D flip-flop has a single input called D (*see* figure 1.5) A D flip-flop is also called a *data* flip-flop because it has one data input. A D flip-flop is also referred to as a *delay* flip-flop as it does not alter the state of the input but merely delays the signal by one clock pulse. This delay operation will be evident if you study the transition table 1.4. If the data input is 1, the output Q is also 1. If the data input is 0, the output Q is also 0. The information (or the input signal) won't be changed, but will be delayed by one clock pulse.

The D flip-flop operates in two modes only: set and reset. When the input data is high R=0 and S=1; therefore the flip-flop is set (Q=1). When the input data is low R=1 and S=0; therefore the flip-flop is reset (Q=0) (*see* table 1.4).

A D Flip Flop is also
called a:
DATA FLIP-FLOP
DELAY FLIP-FLOP
DATA FOLLOWER

what ever the input so will be the output one clock pulse later.

Table 1.4. Transition table for the D flip-flop

Present State Q(t)	D	Next State Q(t+t1)
0	0	0
0	1	1
1	0	0
1	1	1

Note that the D input column is identical to the Q(t+t1) output column.

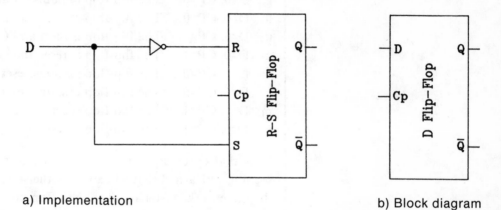

a) Implementation   b) Block diagram

Figure 1.5. D flip-flop

Example

What is the output pulse train of the following D flip-flop? (Use table 1.4)

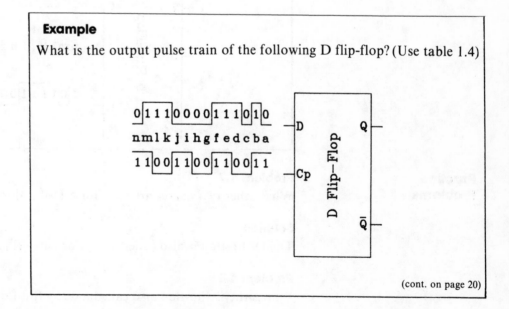

(cont. on page 20)

Solution

a: D=0, CP=1   The flip-flop is reset and Q=0.
b: D=1, CP=1   The flip-flop is set and Q=1.
c: D=0, CP=0   The flip-flop remains set and Q=1.
d: D=1, CP=0   The flip-flop remains set and Q=1.
e: D=1, CP=1   The flip-flop is set and Q=1.
f: D=1, CP=1   The flip-flop is set and Q=1.
g: D=0, CP=0   The flip-flop remains set and Q=1.
h: D=0, CP=0   The flip-flop remains set and Q=1.
i: D=0, CP=1   The flip-flop is reset and Q=0.
j: D=0, CP=1   The flip-flop is reset and Q=0.
k: D=1, CP=0   The flip-flop remains reset and Q=0.
l: D=1, CP=0   The flip-flop remains reset and Q=0.
m: D=1, CP=1   The flip-flop is set and Q=1.
n: D=0, CP=1   The flip-flop is reset and Q=0.

Note that Q remained unchanged whenever CP=0. From the previous section we learned that when CP=0, the state of *any* clocked flip-flop is the same as the last state of the flip-flop under a clock pulse (CP=1).

The output pulse train is:

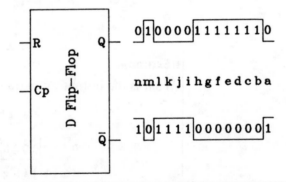

Practice Problems

Problem 1.7
What other two names are used for a D-flip-flop?

Solution
The D-flip-flop is also called a data or delay flip-flop.

Problem 1.8
Why is the D flip-flop also called a delay flip-flop?

Solution
From table 1.4 it is evident that the input signal from the D input is identical to the output Q whenever CP=1. Therefore we can say that the D flip-flop input signal is transferred to the output Q on the presence of a clock pulse. The signal is delayed through the flip-flop. Thus this flip-flop could be operated as a delay

signal when CP=1. The D flip-flop in its delay mode could be used in the synchronization of different components in a sequential circuit.

Problem 1.9

Do past inputs influence the operation of a D flip-flop? Explain!

Solution

The D flip-flop's operation is controlled by one input. If D=1, the D flip-flop is set. When D=0, the D flip-flop is reset. This operation is independent of the flip-flop's state before the pulse. An exception to this application arises when CP=0. Here the D flip-flop ignores the input signal from D, and the last state of the D flip-flop won't change.

Problem 1.10

Obtain the output pulse train of the following D flip-flop:

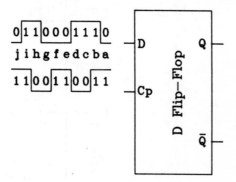

Solution

a:	D=0, CP=1	The flip-flop is reset and Q=0.
b:	D=1, CP=1	The flip-flop is set and Q=1.
c:	D=1, CP=0	The flip-flop remains set and Q=1.
d:	D=1, CP=0	The flip-flop remains set and Q=1.
e:	D=0, CP=1	The flip-flip is reset and Q=0.
f:	D=0, CP=1	The flip-flop is reset and Q=0.
g:	D=0, CP=0	The flip-flop remains reset and Q=0.
h:	D=1, CP=0	The flip-flop remains reset and Q=0.
i:	D=1, CP=1	The flip-flop is set and Q=1.
j:	D=0, CP=1	The flip-flop is reset and Q=0.

The output pulse train is:

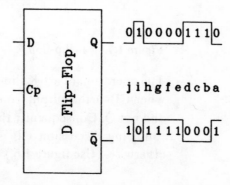

J-K Flip-Flop

(Two input Flip-Flop without a not allowed state.)

In the previous section, we were introduced to a flip-flop with a single input and without a not allowed state. In this section we are going to learn about a two input flip-flop without a not allowed state. When the not allowed state is eliminated, the combinational circuits (or decision making circuits that support the operation of the flip-flop(s) are usually much simpler.) Consquently a two input flip-flop that will accept all combinations of input signals is desirable for design purposes.

A J-K flip-flop is the most universally used flip-flop. A J-K flip-flop is an R-S flip-flop that accepts all combinations of input signals. To implement a J-K flip-flop, two 3-input AND gates are added to an R-S flip-flop as follows:

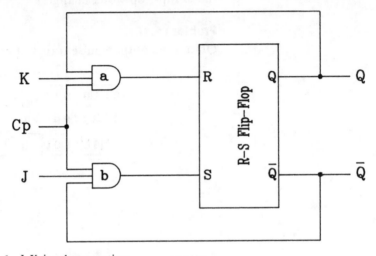

Figure 1.6. J-K implementation

In a J-K flip-flop the S input is referred to as J and the R input is called K (*see* figure 1.7.)

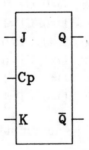

Figure 1.7. J-K flip-flop

The operation of a J-K flip-flop could be obtained using figure 1.6. Note that when CP=0, the output from gates a and b is 0 regardless of the inputs J, K, Q, and not Q. Consequently the R-S flip-flop's state remains unchanged. In the following discussion CP is assumed always high (CP=1) unless specified otherwise. (Use figure 1.6.)

Case 1: Q=0, J=0 and K=0

Inputs to gate a: 0 0 1   (respectively Q K and CP)
Output from gate a: 0

Inputs to gate b: 1 0 1   (respectively $\overline{Q}$ J and CP)
Output from gate b: 0

The output from gate a is the R input to the R-S flip-flop, and the output from gate b is the S input to the R-S flip-flop. We learned previously that when R=S=0, an R-S flip-flop ignores its inputs. Therefore the state of the J-K flip-flop remains unchanged and Q(t)=Q(t+t1)=0 (reset).

Case 2: Q=0, J=0 and K=1

Inputs to gate a: 0 1 1   (Q K CP)
Output from gate a: 0

Inputs to gate b: 1 0 1   ($\overline{Q}$ J CP)
Output from gate b: 0

Since the outputs from gates a and b are 0, R=S=0 and the R-S flip-flop's state is unchanged. The J-K flip-flop is reset and Q(t+t1)=Q(t)=0. (As in case 1.)

Case 3: Q=0, J=1 and K=0

Inputs to gate a: 0 0 1   (Q K CP)
Output from gate a: 0

Inputs to gate b: 1 1 1   ($\overline{Q}$ J CP)
Output from gate b: 1

The output from gate a is 0 and R=0. From gate b the signal is 1, therefore S=1. Previously we learned that when R=0 and S=1, the R-S flip flop is set. Therefore the J-K flip flop is set and Q(t+t1)=1.

Case 4: Q=0, J=1 and K=1

Inputs to gate a: 0 1 1   (Q K CP)
Output from gate a: 0

Inputs to gate b: 1 1 1   ($\overline{Q}$ J CP)
Output from gate b: 1

The outputs from gates a and b are respectively the R and S inputs. When R=0 and S=1 the flip-flop is set. The J-K flip-flop is then set and Q(t+t1)=1. (Same as case 3.)

Case 5: Q=1, J=0 and K=0

Inputs to gate a: 1 0 1 (Q K CP)
Output from gate a: 0

Inputs to gate b: 0 0 1 ($\overline{Q}$ J CP)
Output from gate b: 0

Since R=0 and S=0, the state of the flip-flop remains unchanged and the J-K flip-flop is set (Q(t+t1)=Q(t)=1).

Case 6: Q=1, J=0 and K=1

Inputs to gate a: 1 1 1 (Q K CP)
Output from gate a: 1

Inputs to gate b: 0 0 1 ($\overline{Q}$ J CP)
Output from gate b: 0

Since R=1 and S=0, the flip-flop is reset. The J-K flip-flop is then reset and Q(t+t1)=0.

Case 7: Q=1, J=1 and K=0

Inputs to gate a: 1 0 1 (Q K CP)
Output from gate a: 0

Inputs to gate b: 0 1 1 ($\overline{Q}$ J CP)
Output from gate b: 0

Since R=0 and S=0, the J-K flip-flop's state remains unchanged and Q(t+t)=Q(t)=1. The J-K flip-flop is set.

Case 8: Q=1, J=1 and K=1

Inputs to gate a: 1 1 1 (Q K CP)
Output from gate a: 1

Inputs to gate b: 0 1 1 ($\overline{Q}$ J CP)
Output from gate b: 0

R=1 (output from gate a) and S=0 (output from gate b). The J-K flip-flop is reset and Q(t+t1)=0. The transition table for the J-K flip-flop follows:

Table 1.5. Transition table for a J-K flip-flop

Q(t)	J	K	Q(t+t1)
0	0	0	0
0	0	1	0
0	1	0	1
0	1	1	1
1	0	0	1
1	0	1	0
1	1	0	1
1	1	1	0

We can summarize the operation of a J-K flip-flop as follows:

1 When J=K=0, the state is unchanged.

2 When J=0 and K=1 the J-K flip-flop is reset.

3 When J=1 and K=0 the J-K flip-flop is set.

4 When J=K=1 the output state is complemented. If the flip-flop was set it becomes reset and vice versa.

These four J-K flip-flop operations are known as its *mode of operation*. The J-K flip-flop's modes of operation follow:

1 J=0 K=0, mode: *hold*: Q(t+t1)=Q(t)

2 J=0 K=1, mode: *reset*: Q(t+t1)=0

3 J=1 K=0, mode: *set*: Q(t+t1)=1

4 J=1 K=1, mode: *toggle*: Q(t+t1)=not Q(t)

Example

What is the output train of the following J-K flip-flop? Give the mode of operation for each pulse. (Assume CP=1.)

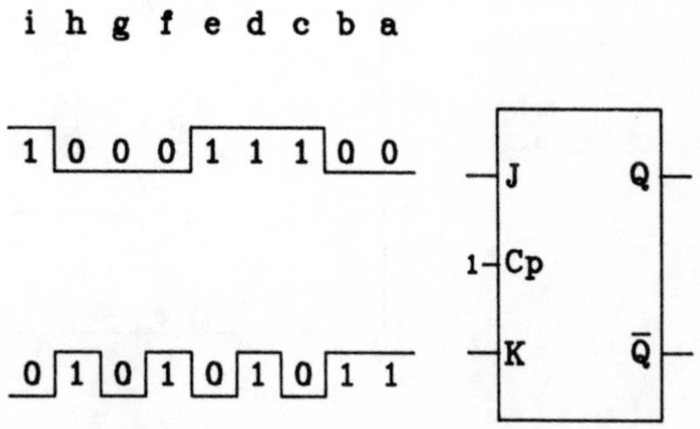

i h g f e d c b a

1 0 0 0 1 1 1 0 0

0 1 0 1 0 1 0 1 1

J   Q

1—Cp

K   Q̄

(cont. on page 26)

Solution

a: J=0, K=0. Reset, Q(t+t1)=0
     Mode: *reset*

b: Reset, same as pulse a
     Mode: *reset*

c: J=1, K=0. Set, Q(t+t1)=1
     Mode: *set*

d: J=K=1.
     The flip-flop complements the output state. After pulse c the flip-flop
     was set. Now the flip-flop is reset and Q(t+t1)=0.
     Mode: *toggle*

e: J=1, K=0. Set, Q(t+t1)=1
     Mode: *set*

f: J=0, K=1. Reset, Q(t+t1)=0
     Mode: *reset*

g: J=K=0.
     The flip-flop remains unchanged. The flip-flop's state is the same as
     in pulse f, Q(t+t1)=Q(t)=0.
     Mode: *hold*

h: J=0, K=1. Reset, Q(t+t1)=0
     Mode: *reset*

i: J=1, K=0. Set, Q(t+t1)=1
     Mode: *set*

The output pulse train is:

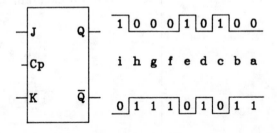

Practice Problems

Problem 1.11
Give the modes of operation of the J-K flip-flop.

Solution
The four modes of operation of a J-K flip-flop are *set, reset, toggle,* and *hold.*

Problem 1.12
Describe the *toggle* and *hold* modes of operations.

Solution
When a J-K flip-flop is under a *hold* mode operation, it ignores its inputs. Its output stays unchanged.

When a J-K flip-flop is under a *toggle* mode it complements its state. In other words its state changes much like a toggle switch.

Problem 1.13
What is the output pulse train of the following J-K flip-flop? For each pulse clock give the mode of operation of the flip-flop. (Assume CP=1.)

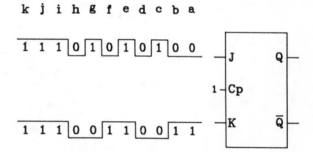

Solution
a   J=0, K=1   *reset*   Q(t+t1)=0
b   J=0, K=1   *reset*   Q(t+t1)=0
c   J=1, K=0   *set*   Q(t+t1)=1
d   J=0, K=0   *hold*   Q(t+t1)=Q(t)=1
e   J=1, K=1   *toggle*   Q(t+t1)=not Q(t)=0
f   J=0, K=1   *reset*   Q(t+t1)=0
g   J=1, K=0   *set*   Q(t+t1)=1
h   J=0, K=0   *hold*   Q(t+t1)=Q(t) =1
i   J=1, K=1   *toggle*   Q(t+t1)=not Q(t)=0
j   J=1, K=1   *toggle*   Q(t+t1)=not Q(t)=1
k   J=1, K=1   *toggle*   Q(t+t1)=not Q(t)=0

Note that pulses i, j and k are all in the toggle mode. Here the state of the J-K flip-flop was toggled from set to reset to set to reset.

When a high pulse is applied to both J and K inputs, the flip-flop will continuously change its state (with CP=1). This phenomena is called oscillation and is usually undesirable.

The output train pulse is:

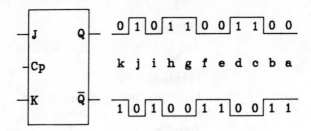

The T Flip-Flop

A T flip-flop (also called a toggle flip-flop) has one input T and two modes of operation: hold and toggle.

When the T input is 0, the mode of operation is hold and Q(t+t1)=Q(t). When T=1, the mode of operation is toggle and Q(t+t1)=not Q(t).

A T flip-flop can be implemented using a J-K flip-flop (*see* figure 1.8).

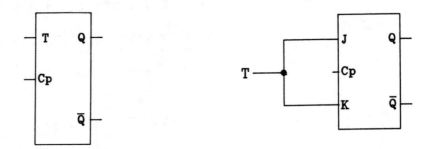

a) Block diagram

b) Circuit implementation

Figure 1.8. T flip-flop

In figure 1.8, we connected the J and K inputs. Once connected the J and K inputs are identical and the flip-flop operates under only two modes: J=K=0 (hold) and J=K=1 (toggle).

Let's summarize the operation of a T flip-flip in the following transition table:

Table 1.6. Transition table for a T flip-flop

(Present state) Q(t)	(Input) T	(Next state) Q(t+t1)
0	0	0
0	1	1
1	0	1
1	1	0

Note that when the input T is 1, the flip-flop's state is complemented; when T is 0, the flip-flop's state remains unchanged.

**Practice
Problems**

Problem 1.14
What are the two modes of operation of a T flip-flop?

Solution
The two modes of operations are: hold and toggle.

Problem 1.15
In what mode of operation is a T flip-flop when its output changes from 1 to 0 to 1 to 0 to 1 to 0...etc.?

Solution
When the output of a T flip-flop changes, it is in the toggle mode.

Problem 1.16
What input signal is required to place the T flip-flop in the toggle mode of operation? What input is required to place it in the hold mode?

Solution
A low level pulse on the T input would put the flip-flop in the hold mode of operation. A high level pulse on the T input would put the flip-flop in the toggle mode of operation.

Problem 1.17
What is the output pulse train of the following T flip-flop? (Assume the flip-flop was reset before the application of the pulse train; also assume CP=1).

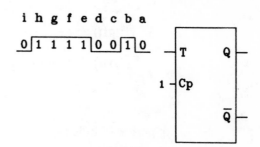

Solution
a   Q(t)=0, T=0   *hold*   Q(t+t1)=Q(t)=0; Q(t)=0 is assumed.
b   Q(t)=0, T=1   *toggle*   Q(t+t1)=not Q(t)=1
c   Q(t)=1, T=0   *hold*   Q(t+t1)=Q(t)=1
d   Q(t)=1, T=0   *hold*   Q(t+t1)=Q(t)=1
e   Q(t)=1, T=1   *toggle*   Q(t+t1)=not Q(t)=0
f   Q(t)=0, T=1   *toggle*   Q(t+t1)=not Q(t)=1
g   Q(t)=1, T=1   *toggle*   Q(t+t1)=not Q(t)=0
h   Q(t)=0, T=1   *toggle*   Q(t+t1)=not Q(t)=1
i   Q(t)=1, T=0   *hold*   Q(t+t1)=Q(t)=1

The output pulse is:

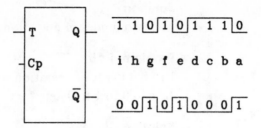

Edge-Triggering of Flip-Flops

Introduction

In the past sections we learned that clocked flip-flops (R-S, D, J-K, and T flip-flops) would respond (transfer signals from input to output) in the presence of a clock pulse. In real applications, flip-flops and the clock have to be *synchronized*. This synchronization involves coordinating clock pulses and input signals, considering timing and delays, and other factors that we will discuss in the following sections.

Definitions

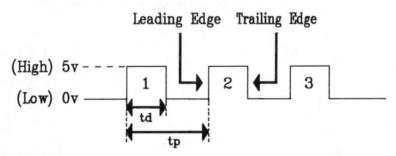

tp: period of pulse
td: duration of pulse

Figure 1.9. Clock pulses for a TTL device

A *clock pulse generator* is a device that generates pulses used in the timing and synchronizing of components. The pulses generated by a clock are voltages. The value of the voltage would depend on the type of device (or IC) to be clocked. For example TTL ICs require 5 volt pulses (*see* figure 1.9): other types of ICs require different voltages.

The pulses generated by a clock generator have to be periodic. The *period* of a clock pulse (tp) is the time required for a pulse to repeat itself (*see* tp on figure 1.9). The *duration* of a clock pulse is the time during which the signal (tor voltage) is high. In general the period of the clock pulse is greater than the duration of a pulse (tp>td).

In figure 1.9 we can see that a pulse is composed of two edges: a leading edge (also called a positive edge) and a trailing edge (also called a negative edge). At the leading edge, the clock switches from low to high (or 0v to 5v). At the trailing edge the clock pulse switches from high to low (or 5v to 0v).

Some flip-flops transfer data from their input(s) to their outputs during the leading edge of a clock; others transfer data on the trailing edge of a clock. Flip-flops that transfer data on the leading edge of a clock are called *positive edge-triggered flip-flops*. Flip flops that transfer data on the trailing edge of a clock are called *negative-edge-triggered flip-flops*. (*See* figure 1.10).

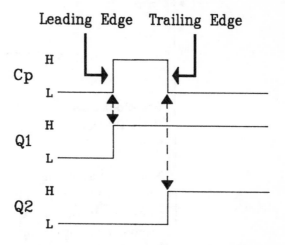

Q1: Output from a positive-edge-triggered flip-flop
Q2: Output from a negative-edge-triggered flip-flop

Figure 1.10. Positive and negative edge triggered flip-flops output

---

Example

Obtain the output pulse train if the following input pulse train were entered into a positive-edge triggered R-S flip-flop.

(cont. on page 32)

Solution

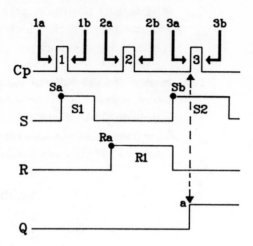

First clock pulse

When the leading edge of the first clock pulse arrives, both inputs (R and S) are low. Therefore the R-S flip-flop output remains unchanged.

Note that Sa (the high signal on the S input) appears after 1a (the leading edge of the first clock pulse) and before 1b (the trailing edge of the first clock pulse). Because this flip-flop is positive-edge-triggered, it won't respond after the leading edge (1a). Therefore this flip-flop ignores input signal S1, because S1 is low when the leading edge (1a) appears.

Second clock pulse

When the leading edge of the second clock pulse (2a) arrives, S is low and the R input is high. The flip-flop is reset and Q=0. Because the flip-flop was reset before the second clock pulse arrived, no change would be observed.

Since Ra (high signal on the R input) appears before 2a (leading edge of the second clock pulse), the flip-flop would respond to this signal.

Third clock pulse

When the leading edge of the third clock pulse arrives, the R input is low and the S input is high. Therefore the flip-flop is set (*see* a at the Q output) and Q=1.

Since Sb occurs before 3a (the leading edge of the third clock pulse), the flip-flop responds to this signal by setting the flip-flop at exactly the same time as when the leading edge of the third clock pulse arrives (*see* 3a and a).

Example

Obtain the output pulse train if the following input pulse train were entered into a negative-edge triggered R-S flip-flop.

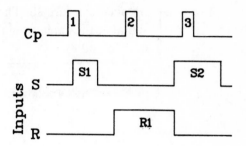

Solution

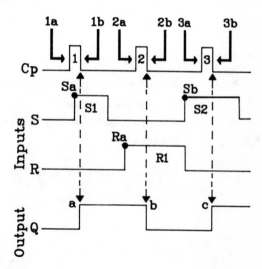

First clock pulse

When the trailing edge of the first clock pulse (1b) arrives, the R input is low and the S input is high. Therefore the flip-flop is set and Q=1 (*see* a).

Sa occurs before 1b. Therefore the flip-flop will respond to the S1 signal exactly when the trailing edge of the first clock pulse arrives (*see* 1b and a).

Second clock pulse

When the trailing edge of the second clock pulse (2b) arrives, the S input is low and the R input is high. Therefore the flip-flop is reset and Q=0 (*see* b).

Ra appears before 2b. Therefore the flip-flop will react to the R1 signal when the trailing edge of the second clock pulse arrives (*see* 2b and b).

(cont. on page 34)

> **Third clock pulse**
>
> When the trailing edge of the third clock pulse (3b) arrives, the R input is low and the S input is high. Therefore the flip-flop is set and Q=1 (*see* c).
>
> Sb occurs before 3b. Therefore the flip-flop will respond to the S2 input signal at the same time the trailing edge of the third clock pulse arrives (*see* 3b and c).

Representation of Positive and Negative-Edge-Triggered Flip-Flops

A positive-edge-triggered flip-flop is represented as follows:

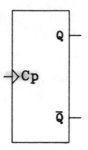

When the symbol $>$ is present at a flip-flop's clock input (R-S, D, J-K, or T), this flip-flop is a positive-edge-triggered flip-flop. Remember that a positive-edge-triggered flip-flop is enabled only in the presence of the leading edge of a clock pulse.

A negative-edge-triggered flip-flop is represented as follows:

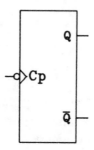

When the symbol o$>$ is present at a flip-flop's clock input, this flip-flop is a negative-edge-triggered flip-flop. Recall that a negative-edge-triggered flip-flop is enabled only in the presence of the trailing edge of a clock pulse.

Throughout this book, all the flip-flops are considered positive-edge-triggered, unless specified otherwise.

Timing

When should the input signal be applied? Should it be applied in sequence with the clock pulse or should it be applied after the clock pulse? In this section we will answer these questions.

In real applications, signals don't switch from low to high or high to low instantly. It takes a certain amount of time for a signal to go from 0 volt to 5 volts and from 5 volts to 0 volt. We say that the signal is fully built when it reaches a stable state (0 or 5 volts). (*See* figure 1.11)

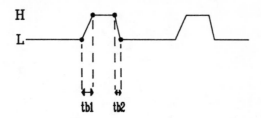

tb1: Time required for a signal to switch from low to high

tb2: Time required for a signal to switch from high to low

Figure 1.11. Stability of signals

Note that tb1 is longer than tb2. In other words, in general, more time is required for a signal to go from low to high, than from high to low.

A device will generally respond to an input signal when 70% of the high or low voltage limit is present. This percentage will vary from one IC manufacturing company to the next. In this book we will assume that IC's respond only when signals are 100% stable. This assumption will eliminate any ambiguities arising from synchronizing IC's. Nonetheless, when designing with IC's you should use a data book to check the timing requirements for a specific IC.

If an input signal is applied to a flip-flop at the same time as a clock pulse, you should ask the following questions:

Is the flip-flop going to respond?

Is the time duration (td) of the clock pulse long enough to allow the input signal to build enough to cause the flip-flop to respond. In other words is td≦tb?

1. $t_d < t_b$

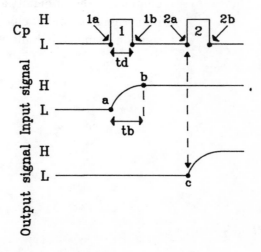

Figure 1.12. Timing diagram with td < tb

Here the clock pulse was not long enough to allow the input signal to be built. We see that the input signal was completely built after pulse 1 (the b is between 1b and 2a). Therefore the flip-flop will only respond after pulse 2 is applied (*see* c).

2. $t_d > t_b$

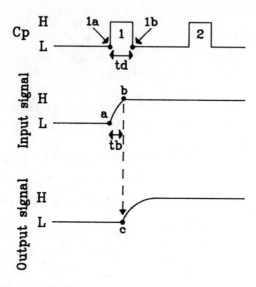

Figure 1.13. Timing diagram with td>tb

The clock pulse was long enough in duration to allow the input signal to be 100% built (**b** is achieved during pulse 1). Therefore the flip-flop responds during the first clock pulse (*see* c).

There is no certain way to determine when a flip-flop starts responding. This may cause some serious problems in the operation of the circuit. One way of eliminating the ambiguity as to when an IC starts responding is by applying the input signal before the clock pulse. By doing so we will be certain that the input signal is completely built when the clock pulse arrives.

Example

Discuss the timing of the following R-S flip-flop:

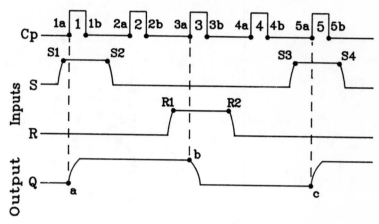

Figure 1.14. R-S flip-flop timing diagram

Solution

When the first clock pulse arrives, the R input signal is low and the S input signal is already 100% built (S1 occurred before 1a). Therefore the R-S flip-flop's output (Q) would start building up exactly when the first clock pulse arrives (*see* a and 1a).

During the second clock pulse both inputs R and S are low. Therefore the R-S flip-flop's output remains unchanged. The R-S flip-flop's output during the second clock pulse is the same as the R-S flip-flop's output during the first clock pulse.

When the third clock pulse arrives, the S input is low and the R input is 100% built (R1 occurs before 3a). Therefore the flip-flop will be reset as soon as the clock pulse arrives (*see* 3a and b).

During the fourth clock pulse, both inputs are low. Therefore the R-S flip-flop's state remains unchanged.

When the fifth clock pulse arrives, the R input is low and the S input is 100% built (S3 occurs before 5a). Therefore the flip-flop will be set and Q will begin building up as soon as the fifth clock pulse arrives (*see* 5a and c).

Example

Complete the output pulse train of the following R-S flip-flop given the S input. Assume that the R input is low and that the flip-flop responds to the presence of a high pulse. Comment on the results!

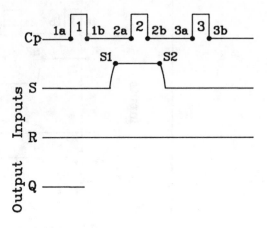

Solution

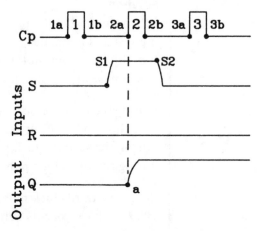

Our analysis will start with the second clock pulse because the output Q was given for the first clock pulse. The input signal S begins building at S1. When the second clock pulse arrives, the input signal S is 100% built and R=0 (given). Therefore the flip-flop would be set and Q would start building at the arrival of the second clock pulse (*see* 2a and a). When the third clock signal arrives, both inputs (R and S) are low. Therefore the output Q would remain unchanged.

Example

Obtain the output pulse train if the following input pulse trains were entered into an R-S flip-flop. (Always assume a positive-edge-triggered flip-flop unless specified otherwise.)

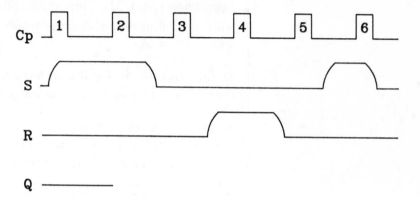

Solution

Remember that a positive-edge-triggered flip-flop is a flip-flop that only responds when the leading edge of clock pulse is present. With that in mind let's determine the output.

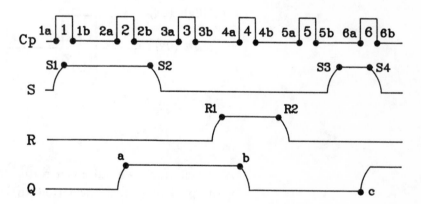

Solution

When the first clock pulse arrives, the R input is low and the S input is not 100% built when the trailing edge of the pulse arrives (S1 occurs after 1a). Therefore the output would remain unchanged and Q=0.

When the second clock pulse arrives, the R input is low and the S input is 100% built (S1 occurs before 2a). Therefore the flip-flop is set and the output Q will start to build up in the leading edge of clock pulse 2 (*see* 2a and a).

(cont. on page 40)

During the third clock pulse both the R and S inputs are low. Therefore the flip-flop's state would remain unchanged from the last clock pulse.

At the fourth clock pulse the S input is low and the R input is high (R1 occurs before 4a). Therefore the flip-flop will be reset and the Q output will start diminishing at the appearance of the leading edge of clock pulse 4 (*see* 4a and b).

During the fifth clock pulse both the R and S inputs are low. Therefore the flip-flop state would remain unchanged from the last clock pulse.

When the sixth clock pulse arrives, the R input is low and the S input is high (S3 occurs before 6a). Therefore the flip-flop is set and the output Q starts to build up at the same moment the leading edge of clock pulse 6 arrives (see 6a and c).

Practice Problems

Problem 1.18

Describe the operation of a master clock. Assume that we're clocking TTL ICs.

Solution

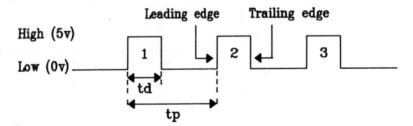

A master clock is a pulse generator. The pulses generated are periodic with a peak of 5 volts. The period (tp) is greater than the duration (td). A pulse is composed of a leading edge (when the voltage rises from 0 to 5 volts) and a trailing edge (when the voltage drops from 5 to 0 volt). Some flip-flops transfer data in the leading edge (or low to high), others in the trailing edge (or high to low) of a clock pulse.

Problem 1.19

Complete the following sentence:

Flip-flops that transfer data from their inputs to their outputs in the positive (or leading) edge of a clock pulse are called ..(a).. Those that transfer data from their inputs to their outputs in the negative (or trailing) edge are called..(b)...

Solution

(a)=positive-edge-triggered flip-flops
(b)=negative-edge-triggered flip-flops

Problem 1.20

When should an input signal be applied? Explain.

Solution

To obtain an appropriate operation, an input signal should be applied before the clock pulse so that the signal is fully built when the clock pulse arrives.

Problem 1.21

Use the following signals to determine if the R-S flip-flops a and b are positive or negative-edge triggered flip-flops.

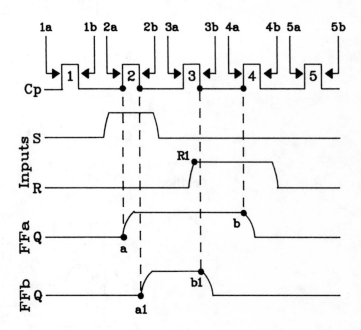

Solution

FF$_a$: The R-S flip-flop is positive-edge-triggered. We can see that from the Q1 signal. At the second clock pulse, the output signal Q1 start building up at the leading edge of pulse 2 (*see* a in the Q1 output). During the third clock pulse, the flip-flop ignores the R signal which was fully built before the trailing edge. This signal is only considered in the leading edge of the fourth clock pulse (*see* b in the Q1 output). We conclude that this flip-flop responds only to the leading edge of a clock pulse. This flip-flop is a positive-edge-triggered flip-flop.

FF$_b$: This R-S flip-flop is negative-edge-triggered. The output of the flip-flop changes in the trailing edge of the clock pulse. a1 corresponds to the trailing edge of pulse 2. b1 corresponds to the trailing edge of pulse 3.

Note that b1 in Q1 precedes b in Q2 due to the following reasons:

1 Signal R1 was not fully built when the leading edge of pulse 3 appeared. R1 occurs after 3a.) Thus a positive-edge-triggered flip-flop won't respond until the next leading edge of a pulse, and Q1 remains unchanged until the leading edge of pulse 4 (or 4a).

2 Signal R1 is fully built when the trailing edge of pulse 3 arrives. (R1 occurs before 3b.) Since Q2 corresponds to a negative-edge-triggered flip-flop, an output change is seen in Q2 during pulse 3 (or at 3b).

Problem 1.22

Draw the output Q of a positive-edge-triggered R-S flip-flop given the following clock and input signals.

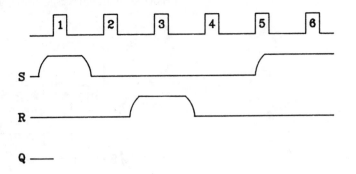

Solution

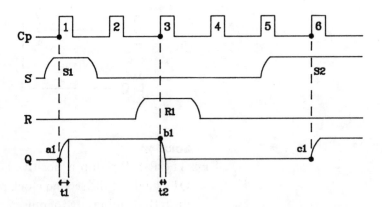

First clock pulse

The input signal S1 is fully built when the leading edge of pulse 1 arrives. Thus the flip-flop's Q output goes high at the same moment that the leading edge of pulse 1 arrives (*see* a1).

Second clock pulse

R and S are both low. The output remains unchanged and Q=1.

Third clock pulse

The input S is low and signal R1 is fully built when pulse 3 arrives. Therefore the R-S flip-flop is reset and the output Q starts decreasing with the leading edge of pulse 3 (see b1).

Fourth clock pulse

R and S are low. The output remains unchanged with Q=0.

Fifth clock pulse

R is low and S is also low. The output remains unchanged and Q=0.

Sixth clock pulse

R is low. S2 is fully built when the leading edge of pulse 6 arrives. Thus the flip-flop is set and Q starts increasing at the leading edge of pulse 6 (*see* c1).

Note that the time required for a signal to switch from low to high is usually longer than the time required for a signal to switch from high to low (compare t1 and t2).

Also note in the fifth clock pulse, the S input signal is built *after* the leading edge of the clock pulse and *before* the trailing edge. Because the flip-flop used is positive-edge-triggered, it won't respond to the signal at the trailing edge. Therefore the output remains unchanged until the next leading edge of the clock pulse.

Problem 1.23

Draw the output Q of a negative-edge-triggered R-S flip-flop given the following clock and input signals.

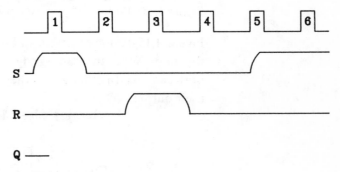

Solution

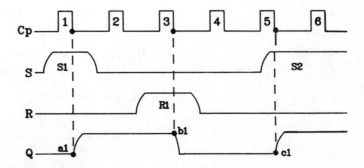

Now the R-S flip-flop responds at the trailing edge of the clock pulse instead of the leading edge of the clock. This is because the flip-flop used in problem 1.22 is negative-edge-triggered. Note that the fifth clock pulse now sets the flip-flop. In problem 1.22 this signal was ignored during pulse 5 because the S input signal is fully built when the *trailing edge* of the clock pulse arrives. If we compare the output signals of problems 1.22 and 1.23, we note that Q's output signals appear at different time intervals. This difference is caused by the two different type of flip-flops used (positive and negative-edge-triggered).

Master-Slave Flip-Flops

Introduction

When outputs of flip-flops are used as inputs to other flip-flops, they (output signals) must be stable and fully built by the time the flip-flops are enabled by the master clock. If not ambiguities may arise.

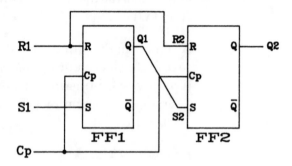

Figure 1.15. Positive-edge triggered R-S flip-flop circuit

Figure 1.15 demonstrates an ambiguity that arises when flip-flops trigger other flip-flops. When the leading edge of the master clock arrives, FF1 and FF2 are both enabled. In other words they are transferring signals from their inputs to their outputs.

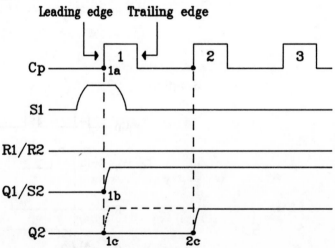

Figure 1.16. Timing of circuit depicted in figure 1.15

The outputs of the second flip-flop are always the same as those of the first flip-flop after one clock pulse. This is why the second flip-flop is called the *slave* and the first flip-flop is called the *master*. The slave will always follow the master. Whenever a certain signal is expected from the output of the first flip-flop, the same signal will eminate from the second flip-flop's output.

Referring to figure 1.17, note that the output Q1 of the first flip-flop is also the S input to the second flip-flop and output Q2 of the first flip-flop is the R input to the second flip-flop. Therefore when Q1=0 and Q2=1 we have R2=1 and S2=0. The second flip-flop (the slave) is reset so Q3=3 and Q4=1. If Q1=1 then Q2=0, and we have R2=0 and S2=1. The slave is set and Q3=1 and Q4=0. We conclude that Q1=Q3 and Q2=Q4 is true after a full clock pulse. (*See* figure 1.17).

Suppose now that both flip-flops of figure 1.17 are positive edge triggered flip-flops. In other words these flip-flops transfer data from their inputs to their outputs only during the leading edge of a clock pulse. Now examine their CP input.

A NOT gate is used to invert the clock signal in the second R-S flip-flop (or slave). In other words when the clock from the master clock is high, the second flip-flop (or slave) will see it low. Conversely when the signal from the master clock is low the second flip-flop will see it high. We can say that when a leading edge appears in the first flip-flop (master), a trailing edge will appear in the second flip-flop (slave) and vice versa. Thus when the master is enabled, the slave is disabled and vice versa.

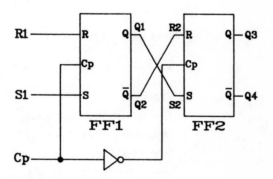

Figure 1.17. Master-slave R-S flip-flop

In figure 1.17, the master is enabled with the leading edge of the master clock, and the slave is enabled with the trailing edge of the same clock pulse. If an input signal(s) is applied to the first flip-flop (or master), this signal will appear at Q1 and its complement at Q2 in the leading edge of the clock pulse. At the trailing edge of the same clock pulse, the signals on Q1 and Q2 will be transferred through flip-flop 2 (the slave) and will appear at Q3 and Q4. Recall that Q1=Q3 and Q2=Q4. We've used the full clock pulse (positive and negative edges of the pulse) to achieve one transfer of data. The outputs from the master-slave flip-flop only appear in the trailing edge (which disables the inputs) of the master clock. Therefore the outputs that are going to be used as inputs to other component(s) would be fully built when the next leading edge of the master clock arrives.

Figure 1.18 represents the clock signal seen by the master slave flip-flop. Note that the clock signal seen by the slave is the complement of the clock signal seen by the master, this is caused by the NOT gate. Notice that td is longer for the slave. Therefore the output signal would have enough time to build before the leading edge of the next clock pulse arrives.

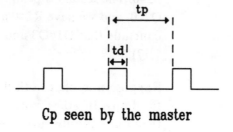

Cp seen by the master

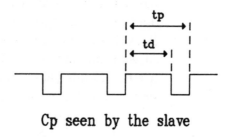

Cp seen by the slave

Figure 1.18. Clock pulses to the Master and Slave.

Example

Complete the following timing diagram. Describe step by step the operation of the following master-slave R-S flip-flop. Assume the master-slave was reset before the first clock pulse. (FF1 and FF2 are positive edge-triggered.)

(cont. on page 47)

Solution

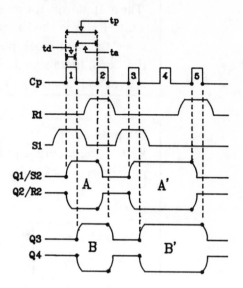

Before 1st clock pulse

Master slave is reset. Therefore Q3=0 and Q4=1.

1st clock pulse

Leading edge (⌐): FF1 is set, Q1=1 and Q2=0 (FF2 disabled)
Trailing edge (⌐): FF2 is set, Q3=1 and Q4=0 (FF1 disabled)

2nd clock pulse

Leading edge (⌐): FF1 is reset, Q1=0 and Q2=1 (FF2 disabled)
Trailing edge (⌐): FF2 is reset, Q3=0 and Q4=1 (FF1 disabled)

3rd clock pulse

Leading edge (⌐): FF1 is set, Q1=1 and Q2=0 (FF2 disabled)
Trailing edge (⌐): FF2 is set, Q3=1 and Q4=0 (FF1 disabled)

4th clock pulse

Leading edge (⌐): FF1 unchanged, Q1=1 and Q2=0 (FF2 disabled)
Trailing edge (⌐): FF2 is set, Q3=1 and Q4=0 (FF1 disabled)

5th clock pulse

Leading edge (⌐): FF1 is reset, Q1=0 and Q2=1 (FF2 disabled)
Trailing edge (⌐): FF2 is reset, Q3=0 and Q4=1 (FF1 disabled)

Note that the shift between A, A' and B B' is equal to td (duration of one clock pulse).

The output of the master-slave (Q3 and Q4) appears at the trailing edge of a pulse. Therefore it takes one clock pulse (both edges) to transfer data from inputs (R and S) to outputs (Q3 and Q4).

(cont. on page 48)

The output from the master-slave has an available time ta (ta=tp-td) to be built. Refer to the figure on page 47. Therefore it would be 100% built when the next clock pulse arrives. If Q3 and Q4 are inputs to other flip-flops, these signals are fully built by the time the next clock pulse arrives and no ambiguity is raised (*see* next example).

Example

Repeat the previous example for the following circuit, using only the initial three clock pulses.

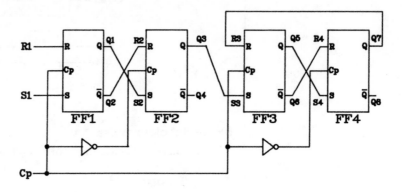

Master Slave I: FF1 and FF2
Master Slave II: FF3 and FF4

Solution

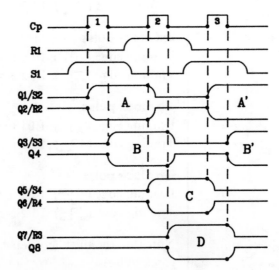

Before 1st clock pulse

All the flip-flops are reset. Therefore Q1=Q3=Q5=Q7=0 and
Q2=Q4=Q6=Q8=1.

(cont. on page 49)

1st clock pulse

Leading edge

FF2 and FF4 are disabled. FF1 is set (R1=0, S1=1). Therefore Q1=1 and Q2=0. FF3 is unchanged. (Q7=R3=0 and Q3=S3=0). Therefore Q5=0 and Q6=1

Trailing edge

FF1 and FF3 are disabled. FF2 is set. (Q1=S2=1, Q2=R2=0). Therefore Q3=1 and Q4=0. FF4 is reset. (Q5=S4=0 and Q6=R4=1). Therefore Q7=0 and Q8=1

2nd clock pulse

Leading edge

FF2 and FF4 are disabled. FF1 is reset. (R1=1, S1=0). Therefore Q1=0 and Q2=1. FF3 is set. (Q3=S3=1, Q7=R3=0). Therefore Q5=1 and Q6=0.

Trailing edge

FF1 and FF3 are disabled. FF2 is reset. (Q1=S2=0, Q2=R2=1). Therefore Q3=0 and Q4=1. FF4 is set. (Q5=S4=1, Q6=R4=0). Therefore Q7=1 and Q8=0.

3rd clock pulse

Leading edge

FF2 and FF4 are disabled. FF1 is set. (R1=0, S1=1). Therefore Q1=1 and Q2=0. FF3 is reset. (Q7=R3=1, Q3=S3=0). Therefore Q5=0 and Q6=1.

Trailing edge

FF1 and FF3 are disabled. FF2 is set. (Q1=S2=1, Q2=R2=0). Therefore Q3=1 and Q4=0. FF4 is reset. (Q5=S4=0, Q6=R4=1). Therefore Q7=0 and Q8=1.

Note that when the master-slave II (more precisely FF3) is clocked (leading edge), its input signals (Q7 and Q3) are fully built. (refer to the figure). These input signals could not be partially built when the leading edge of a master clock arrives because they are outputs of slaves (FF2 and FF4) which are triggered in the trailing edge.

**Other Master-
Slave Flip-Flops**

**J-K Master-Slave
Flip-Flops**

Any flip-flop (J-K, D or T) can be used as a master-slave implementation. We'll example these various master-slave implementations.

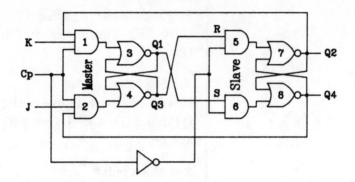

Figure 1.19. A J-K master-slave flip-flop

The timing of the J-K master-slave (as well as all the other master-slaves) is identical to the R-S master-slave. The differences between the master-slaves lies in their operation, *and* not in their control (or triggering) as we'll see in the practice problems.)

**D and T
Master-Slave
Flip-Flops**

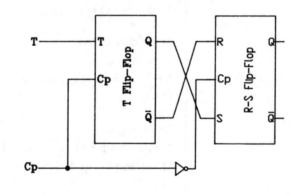

a) T

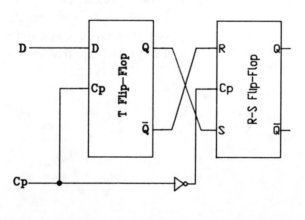

b) D

Figure 1.20. Master-slave flip-flops

Practice Problems

Problem 1.24

Discuss the advantages of a master-slave configuration.

Solution

The principal advantage of a master-slave configuration is its timing. A master-slave flip-flop uses a full clock signal to transfer data from its inputs to its outputs. Very often the outputs of flip-flops are used as inputs to other flip-flops and so on. The outputs from the flip-flops must be fully built by the time the flip-flops are enabled by the master clock so that an appropriate operation can be obtained. When using a master-slave configuration, the outputs have the time necessary to build (tp-td) so that when the next clock pulse arrives, the signals are fully built and an appropriate operation is obtained.

Problem 1.25

Discuss the role of the slave.

Solution

The slave is a flip-flop that is clocked so that it is enabled by the opposite clock edge which enables the master. The major role of the slave is to delay the output signal until the flip-flop's inputs are disabled. This prevents the flip-flops (masters) from accepting input signals until they are fully built.

The slave's output state is always identical to the master's output state. Therefore the slave's output state tells what bit is stored by the master-slave. For example if the output Q of a master-slave R-S flip-flop is 0, then the flip-flop is storing binary bit 0. If the output is 1 the flip-flop is storing binary bit 1.

Problem 1.26

Given the following master clock signal, determine the clock signal that is seen by the slave.

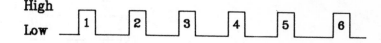

Solution

The clock signal input to the slave is:

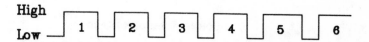

Note that the clock signal input to the slave is always the complement of that of the master clock. This is caused by the NOT gate on the slave clock input.

Problem 1.27

Complete the following timing diagram of an R-S flip-flop that is positive-edge triggered.

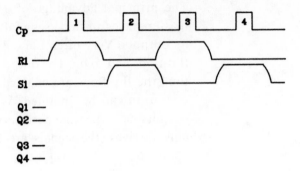

Solution

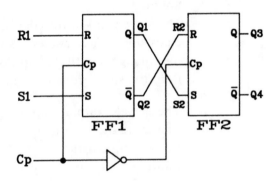

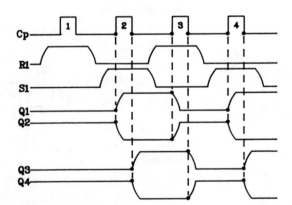

Before 1st clock pulse

The master and slave are both reset. Therefore Q1=Q3=0 and Q2=Q4=1.

After 1st clock pulse

The R-S master-slave flip-flop is reset and Q3=0.

After 2nd clock pulse

The R-S master-slave flip-flop is set and Q3=1.

After 3rd clock pulse

R-S master-slave flip-flop is reset and Q3=0.

After 4th clock pulse

R-S master-slave flip-flop is set and Q3=1.

Problem 1.28

Repeat problem 1.27 for the following timing diagram.

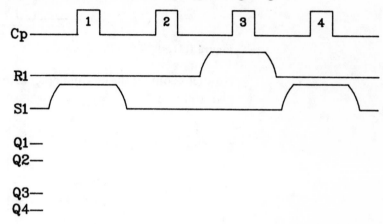

Solution

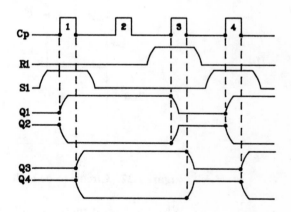

Experiment 1

Purpose

Demonstrate the operation of a NOR R-S flip-flop.

Parts List

Quantity	Parts Description
1	Temporary board
1	Power supply (5v)
2	200 ohm resistors
2	Leds
2	Switches
1	7402 IC (NOR gates)

**Circuit
Implementation**

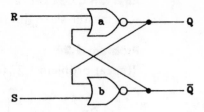

Figure 1.21. R-S flip-flop

Now let's convert the circuit depicted in figure 1.21 into the practical circuit we plan to build.

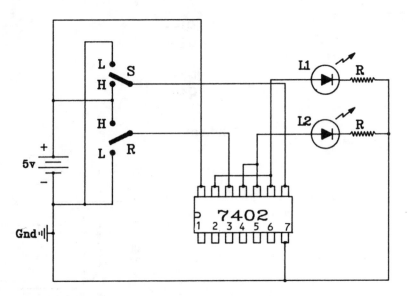

Figure 1.22. Circuit to be built.

Wire the circuit pictured in figure 1.22. Test the circuit as follows:

Use switches S and R to input signals into the R-S flip-flop. If the S and R switches are low, then the inputs to the R-S flip-flop are 0 0. If the S switch is in the high position and the R switch is in the low position, the inputs to the R-S flip-flop are 0 (R) and 1 (S). If the switch S is low and if the R switch is high, the inputs are 1(R) and 0(S).

The outputs Q and not Q are represented by L1 and L2 respectively. When an led is on, it represents binary 1. When an led is off it represents binary 0.

Test all the possible input combinations and record the results in table 1.7. Compare table 1.7 with table 1.1. They should be identical.

Table 1.7. R-S flip-flop transition table

Q(t)	S	R	Q(t+t1)	$\overline{Q}$(t+t1)
0	0	0		
0	0	1		
0	1	0		
0	1	1		
1	0	0		
1	0	1		
1	1	0		
1	1	1		

Note that Q(t) represents the output of the R-S flip-flop before the application of the R and S signals. Q(t+t1) and not Q(t+t1) depict the output of the R-S flip-flop after the input signals have been applied.

Experiment 2

Purpose

Demonstrate the operation of a D type flip-flop

Parts List

Quantity	Parts Description
1	Temporary board
1	Power supply (5v)
2	200 ohm resistors
2	Leds
2	Switches
1	7475 IC (quad D flip-flops)

**Circuit
Implementation**

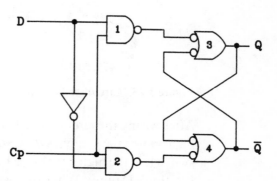

Figure 1.23. D flip-flop implementation

The 7475 IC is a quad D type flip-flop. In other words, it contains four D flip-flops. The T input is equivalent to the clock pulse input or CP. In our experiment we will wire and test one D flip-flop. We'll be using D flip-flop #4 in the experiment.

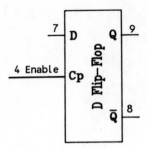

Figure 1.24. D flip-flop

Now let's experiment with the 7475 by wiring the circuit depicted in figure 1.25. (Refer to the earlier experiments for wiring hints).

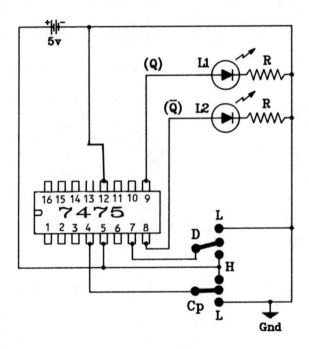

Figure 1.25. Circuit to be built

Testing the D Flip-Flop

After wiring the circuit depicted in figure 1.25 switch the CP input switch to high, switch the D input switch to low and observe L1 and L2. When the D inut switch is in the high position, a high level input signal (or 1) is applied to the D flip-flop's D input. When the switch is in the low position, a 0 is applied to the D flip-flop's D input. The state of the flip-flop can be seen from L1 and L2. When L1 is on, the flip-flop is set and a binary 1 is stored in the D flip-flop. When L1 is off, the flip-flop is reset and a 0 is stored in the D flip-flop. Note that when L1 is on, L2 is off and vice versa.

Complete the testing and record the results in the following table. If the led is on record a 1; if it is off, record a 0.

Table 1.8. D-flip-flop transition table

Q(t)	D	Q(t+t1)
0	0	
0	1	
1	0	
1	1	

Q(t): State of flip-flop before applying the input pulse (D).
Q(t+t1): State of flip-flop after applying the input pulse (D).

Results Tables 1.8 and 1.4 are identical.

Experiment 3

Purpose Demonstrate the operation of a master-slave J-K flip-flop

Parts List

Quantity	Parts Description
1	Temporary board
1	Power supply (5v)
2	200 ohm resistors
2	Leds
5	Switches
1	7476 IC (dual J-K flip-flop)

Circuit Implementation

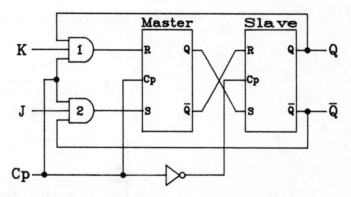

Figure 1.26. Basic J-K flip-flop

Figure 1.26 represents a basic J-K flip-flop. Additional circuitry is generally added to the J-K flip-flop of figure 1.26 to make it easier to use and interface. The most important additional features are a set and reset pin. These pins force the J-K flip-flop to reset or set no matter what the inputs (J, K, and CP) are. In other words these pins override the other input pins. Figure 1.27 shows an implementation of a J-K flip-flop including these new features.

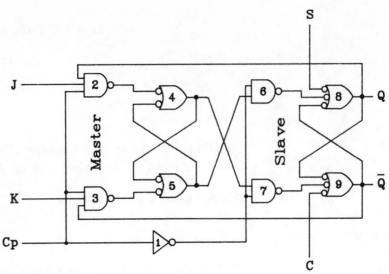

Figure 1.27. Advanced master-slave J-K flip-flop

The C input stands for clear (or reset), and the S input stands for set. Both the C and S inputs are low true. In other words they are activated by a low signal. When a low signal (or 0) is applied to the S input, a high level pulse (or 1) appears at the output of gate 8. The flip-flop is set (Q=1). If a low signal (or 0) is applied to the C input, a high level signal (or 1) appears at the output of gate 9. This high signal from gate 9 will trigger gate 8 and a low level signal (or 0) appears at the output of gate 8. The flip-flop is reset and Q=0. Note that C and S are inputs to gates 8 and 9 which directly control the state of the flip-flop (Q is the output of gate 8 and not Q is the output of gate 9.) For this reason these inputs (S and C) are called asynchronous inputs. By asynchronous we mean that their operation is independent of the master clock and from the J and K inputs. The asynchronous mode of operation (when using the C and S inputs) overrides the synchronous mode of operation (when using J and K inputs).

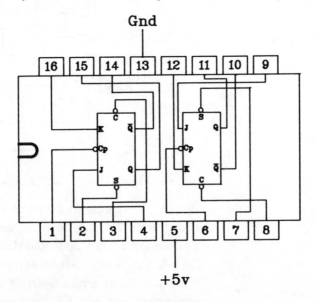

Figure 1.28. Dual master-slave J-K flip-flop (7476 IC)

Figure 1.28 represents a commercially available dual master-slave J-K flip-flop used in our experiments.

Now wire the circuit of figure 1.29.

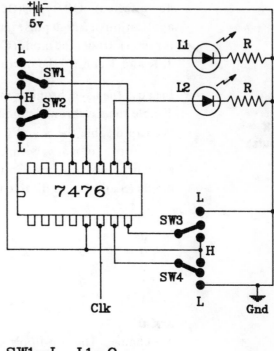

SW1=J   L1=Q

SW2=K   L2=$\overline{Q}$

SW3=C

SW4=S

Figure 1.29. 7476 IC test

**Test Procedure -
Asynchronous
Operation
(C and S Inputs)**

Ignore the T, J and K inputs and complete the following table. (Remember that if an led is on, 1 should be recorded in the table. If an led is off, 0 should be recorded).

Inputs		Outputs			
S	C	Q		$\overline{Q}$	
0	0	...	(a)	...	(b)
0	1	...	(c)	...	(d)
1	0	...	(e)	...	(f)
1	1	...	(g)	...	(h)

Test Results

S=C=0

An ambiguous state results and Q=$\overline{Q}$=1 (a=b=1).

S=0, C=1

The flip-flop is set and Q=1 (c=1 and d=0).

S=1, C=0

The flip-flop is reset and Q=0 (e=0 and f=1).

S=C=1

No change. (The asynchronous inputs are disabled and the flip-flop's state is not affected). Q(t+t1)=Q(t). If the flip-flop was set (or reset) before the application of a high pulse into the S and C inputs, the flip-flop will remain set (or reset). Under the mode (C=S=1) the flip-flip is in the synchronous mode and J, K and T control the operation of the flip-flop.

Test Procedure - Synchronous Operation (J, K and Clock Inputs)

Because the asynchronous mode overrides the synchronous one, we have to disable the asynchronous mode to obtain an appropriate synchronous mode. We can disable the S and C inputs by applying a high level pulse to both. Therefore, to work with the synchronous mode, start by setting the asynchronous switches (C and S switches) to high. Once the asynchronous inputs (S and C) are disabled, complete the following table:

Inputs		Outputs			
J	K	Q		$\overline{Q}$	
0	0	...	(a)	...	(b)
0	1	...	(c)	...	(d)
1	0	...	(e)	...	(f)
1	1	...	(g)	...	(h)

Results

J=K=0

No change. The flip-flop's state remains unchanged after the clock pulse. Q(t)=Q(t+t1).

J=0 K=1

The flip-flop is reset. Q=0 (c=0 and d=1).

J=1 K=0

The flip-flop is set. Q=1 (e=1 and f=0).

J=K=1

Toggle mode. The flip-flop's output Q is complemented every time a clock pulse is applied.

Problems

Problem 1.1

Define a sequential circuit.

Problem 1.2

What is the output pulse Q of the following R-S flip-flop?

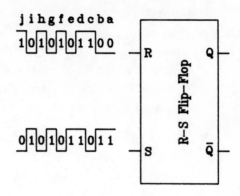

Problem 1.3

What is the output pulse Q of the following R-S flip-flop?

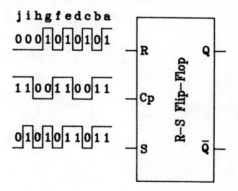

Problem 1.4

What is the output pulse Q of the following D flip-flop?

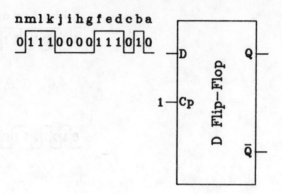

Problem 1.5

What is the output pulse of the following J-K flip-flop?

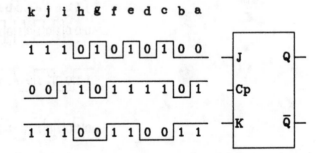

Problem 1.6

What is the output pulse Q of the following T flip-flop? Assume the flip-flop is set.

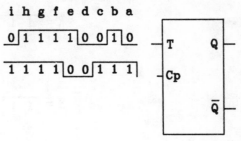

Problem 1.7

Obtain a D flip-flop using an R-S flip-flop.

Problem 1.8

Obtain a T flip-flop using a J-K flip-flop.

Problem 1.9

Obtain the output pulse if the following input pulses were entered into a positive-edge-triggered R-S flip-flop.

$$R = \overline{1 \quad 1 \quad 1} \lfloor 0 \rfloor \overline{1} \lfloor 0 \rfloor \overline{1} \lfloor 0 \rfloor \overline{1} \lfloor 0 \quad 0$$

$$S = \overline{1 \quad 1 \quad 1} \lfloor 0 \quad 0 \rfloor \overline{1 \quad 1} \lfloor 0 \quad 0 \rfloor \overline{1 \quad 1}$$

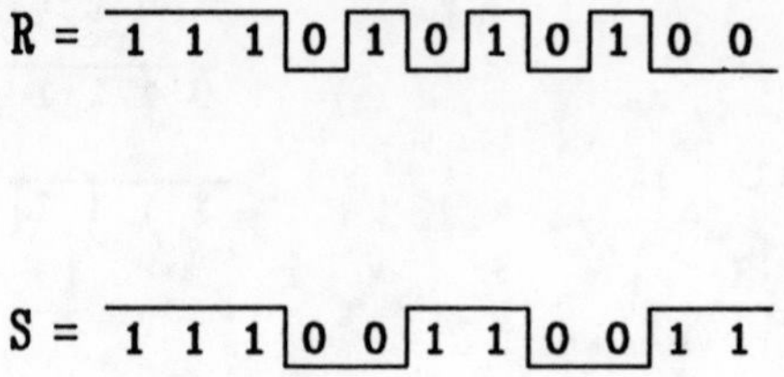

Problem 1.10.

Repeat problem 1.9 for a negative-edge-triggered R-S flip-flop.

Introduction

Counters are sequential logic circuits that count the number of pulses applied to them. Counters are made up of a number of flip-flops as shown in figure 2.1.

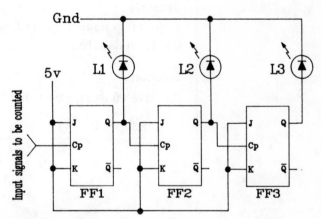

Figure 2.1. Modulo 8 binary counter

The number of flip-flops that a counter contains determines its maximum count or *modulo*. A modulo 4 counter is a counter that can count from binary 00 to 11 or from decimal 0 to 3. A modulo 16 counter counts from binary 0000 to 1111 or from decimal 0 to 15. Once the counter arrives at its maximum count it is reset and starts to count again. For example, a modulo 8 counter will count as follows:

Table 2.1. Binary count for a modulo 8 counter

Binary count			Decimal
L3	L2	L1	equivalent
0	0	0	0
0	0	1	1
0	1	0	2
0	1	1	3
1	0	0	4
1	0	1	5
1	1	0	6
1	1	1	7

The modulo of a counter can be determined by the number of flip-flops. Suppose n represents the number of flip-flops and *mod* is the modulo of a counter; we can write:

$$mod = 2^n \qquad (2.1a)$$

Conversely, from the modulo we can derive the number of flip-flops, by transforming equation 2.1a as follows:

$$
\begin{aligned}
mod &= 2^n \\
\ln(mod) &= \ln(2^n) \\
\ln(mod) &= n * \ln(2) \\
\ln(mod)/\ln(2) &= n \qquad (2.1b)
\end{aligned}
$$

Where "ln" represents the logarithm Neperien (*see* example).

Example

Obtain the number of flip-flops needed to design a modulo 32 counter. Use formula 2.1b.

Solution

We have mod=32; now using 2.1b we arrive at:

$\ln(32)/\ln(2) = 5$ flip-flops

**Logarithm
Neperien
Computations**

Figure 2.2. Calculator designed to evaluate a logarithm neperien function

Figure 2.2 depicts a calculator that is designed to evaluate a logarithm neperien function. Observe the following steps to compute ln(32) and ln(2):

1 Press 3 and then 2 to enter the number 32.

2 Press the ln key to compute ln(32).
The result displayed is 3.465735903.

3 Press the division button.

4 Press 2 to enter the number 2.

5 Press the ln key to compute ln(2).
The result displayed is 0.69314718

6 Press the equal button to compute ln(32)/ln(2)
The final result displayed is 5.

**Logarithm Neperien
Fractional Values**

If ln(mod)/ln(2) is a number containing a fraction, then add 1 and ignore the fraction.

> **Example**
> How many flip-flops are needed to obtain a counter mod 11?
>
> **Solution**
>   ln(11)/ln(2) = 3.459431619
>
> The number of flip-flops needed is 4.

Four flip-flops allow 2^4 or 16 counts. In the following sections we will see how 4 flip-flops could be used to count to decimal 10, and then be reset to count again from decimal 0.

There are many different types of counters which we will describe in the following sections. For instance, counters can be defined as either up or down. An up counter counts by incrementing as shown below:

FF2	FF1	Decimal equivalent
0	0	0
0	1	1
1	0	2
1	1	3

A down counter counts by decrementing as shown below:

FF2	FF1	Decimal equivalent
1	1	3
1	0	2
0	1	1
0	0	0

Counters can also be classified as asynchronous or synchronous. An asynchronous counter is depicted in figure 2.1. Notice that the output (Q) from FF1 is the clock signal for FF2. Likewise, the clock for FF3 is the output (Q) from FF2. The clock to FF1 is the input (CP) which can be any signal which is to be counted. With synchronous counters, all the flip-flops are triggered by the same input signal. We'll discuss synchronous counters in more detail later in the chapter.

Binary Counters

Counter Operation

A binary counter is made of a number of J-K flip-flops. The flip-flops are cascaded so that a count can be obtained. Before we start explaining counter operation let's refresh our memory by reviewing the modes of operation for a J-K flip-flop.

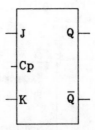

Figure 2.3. J-K flip-flop

Table 2.2. Mode of operation-J-K flip-flop

Inputs		Output	Mode	Description
J	K	Q		
0	0	unchanged	hold	The output of the flip-flop is the same as before the clock pulse.
0	1	0	reset	The flip-flop is reset.
1	0	1	set	The flip-flop is set.
1	1	complement	toggle	The flip-flop's output is complemented each time a clock pulse is applied.

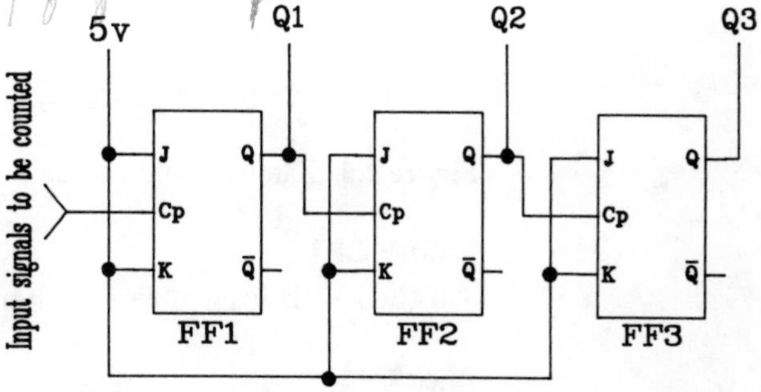

Figure 2.4. Modulo 8 binary counter

Figure 2.4 represents a modulo 8 counter. Note that the three flip-flops are in the toggle mode, because a 5-volt (or 1 = high) signal is applied to the J and K inputs (*see* table 2.2).

Since the flip-flops are in the toggle mode, whenever an input signal arrives at the clock (CP), the flip-flop(s) will complement its state. For example, if the flip-flop's output was 0 before the clock signal arrives, it becomes 1 after the first clock pulse, 0 after the second clock pulse, 1 after the third clock pulse, and so on.

Now examine the clock inputs of FF2 and FF3. The clock input signals are fed from the outputs Q1 and Q2 respectively. Therefore, when FF1 toggles, its output will be complemented (from L to H or H to L), and FF2 may be toggled.

We can say that FF2 and FF3 are independent from the master clock since they receive their clock signals from Q1 and Q2 instead of the master clock. Consequently the operation of this counter is asynchronous.

Now let's analyze the operation of the modulo 8 counter from its timing diagram (*see* figure 2.5). Assume all the flip-flops are reset before the application of the first input signal.

Remember that FF1, FF2, and FF3 are in the toggle mode (J and K are kept high). Note also that the flip-flops toggle only when the high to low transition occurs. We can therefore assume these are negative edge-triggered.

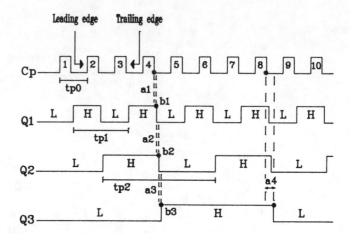

Figure 2.5. Timing diagram for a modulo 8 counter

Before CP1
All the flip-flops are reset. Therefore Q3 = Q2 = Q1 = 0.

After CP1
At the high to low clock transition (or trailing edge) of CP1, FF1 is set and Q1=1. Q1 has gone from low to high; this transition does not toggle FF2. Therefore Q2 = 0. Since there is no change in FF2, consequently there is no change in FF3 and Q3=0.

The outputs after CP1 are Q3 Q2 Q1 = 0 0 1.

After CP2
At the high to low transition of CP2, FF1 is reset and Q1=0. Q1 has gone from high to low. This high to low transition toggles FF2. FF2 is set and Q2=1. Q2 has gone from low to high. This won't toggle FF3. FF3 remains reset and Q3=0.

The outputs after CP2 are Q3 Q2 Q1 = 0 1 0.

After CP3
At the high to low transition of CP3, FF1 is set and Q1=1. Q1 has gone from low to high; this low to high transition won't toggle FF2 and FF2 remains unchanged. Therefore Q2=1. Because FF2 remains unchanged, FF3 is also unchanged and Q3=0.

The outputs after CP3 are Q3 Q2 Q1 = 0 1 1.

After CP4

The high to low transition of the clock resets FF1. Therefore Q1=0. Q1 has gone from high to low; this high to low transition toggles FF2. FF2 is then reset and Q2=0. Q2 has gone from high to low; this transition (high to low) toggles FF3. FF3 is then set and Q3=1.

The outputs after CP4 are Q3 Q2 Q1 = 1 0 0.

After CP5

The high to low transition of the clock sets FF1. Therefore Q1=1. Q1 has gone from low to high. This won't cause FF2 to toggle and Q2 remains unchanged. Thus Q2=0. Since there was no change on FF2, FF3 remains unchanged and Q3=1.

The outputs after CP5 are Q3 Q2 Q1 = 1 0 1.

After CP6

The transition high to low of CP6 causes FF1 to be reset. Therefore Q1=0. Q1 has gone from high to low; this transition causes FF2 to toggle. Thus FF2 is set and Q2=1. Q2 has gone from low to high; this transition won't toggle FF3. Therefore FF3 remains unchanged and Q3=1.

The outputs after CP6 are Q3 Q2 Q1 = 1 1 0.

After CP7

The high to low transition of CP7 causes FF1 to be set. Therefore Q1=1. Q1 has gone from low (reset) to high (set). The transition of FF1 (low to high) won't toggle FF2. Therefore Q2=1. There is no change in FF2. Consequently FF3 remains unchanged, and Q3=1.

The outputs after CP7 are Q3 Q2 Q1 = 1 1 1.

After CP8

The high to low transition of CP8 causes FF1 to be reset and Q1=0. Q1 has gone from high (set) to low (reset). This transition of Q1 will cause FF2 to toggle. FF2 is reset and Q2=0. Q2 has gone from high (set) to low (reset). This transition of Q2 will cause FF3 to toggle. FF3 is reset and Q3 = 0. The three flip-flops are reset and the count starts at 0 0 0 again.

The outputs after CP8 are Q3 Q2 Q1 = 0 0 0.

After CP9

Same as after CP1.

The outputs after CP9 are Q3 Q2 Q1 = 0 0 1.

After CP10

Same as after CP2

The outputs after CP10 are Q3 Q2 Q1 = 0 1 0.

FF1 toggles every time a high to low transition occurs on the clock pulse. FF2 toggles every time a high to low transition occurs at the output Q1 of FF1, or whenever FF1 is reset. Finally FF3 is toggled when FF2's output (Q2) goes from high to low (or FF2 is reset).

When counting, FF3's output represents the MSB (most significant bit), and FF1's output represents the LSB (least significant bit) of the binary count.

Frequency and Counters

Counters are often used as frequency dividers. A frequency is the rate of occurrence of a signal. The frequency is measured in Hertz (Hz). *The frequency is the inverse of the period.* The period is the time (in seconds) that a signal takes to repeat itself.

Example

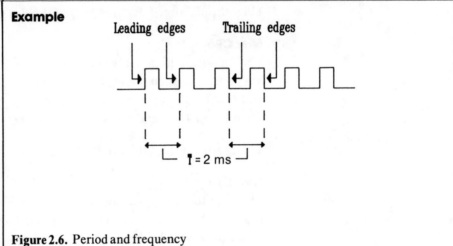

Figure 2.6. Period and frequency

From figure 2.6 we have:

Period = T = 2 ms = 2E–3 s
Frequency = F = 1/2ms = 1/2 E–3 = 500 Hz

In figure 2.5 we see that the period of Q2 (tp2) is twice that of Q1 (tp1) and that the period of Q1 (tp1) is twice that of the clock (tp0). We can write:

$$tp2 = 2 * tp1 = 4 * tp0 \tag{2.2a}$$

or

$$tp0 = tp1 / 2 = tp2 / 4 \tag{2.2b}$$

Use the following formula to convert from period to frequency:

$$\text{Period (tp)} = 1 / \text{frequency (f)} \tag{2.2c}$$

where the period (tp) is in seconds (s) and the frequency (f) is in hertz (hz)

Now

$$f0 = 2 * f1 = 4 * f2 \qquad (2.2d)$$

From 2.2d we see that each flip-flop divides its input frequency by 2. If the frequency of the master clock was 6 Mhz, the frequency at Q1 is 3 Mhz. At Q2 the frequency is 1.5 Mhz (3/2), and at Q3 (FF3) the frequency is 0.75 Mhz. We can write:

$$f(Qn) = (1 / 2^n) * f(CP) \qquad (2.2e)$$

where

Qn   output number (for example at Q5 we have n=5)
f(Qn)   frequency at output Qn
f(CP)   frequency of the clock master

Example

If the frequency of the master clock f(CP) is 2khz, what is the frequency at Q1, Q2 and Q3?

Solution

Using 2.2e we obtain

   a) frequency at Q1 = f(Q1)

      $f(Q1) = (1 / 2^1) * f(CP)$
      $f(Q1) = (1/2) * 2khz$
      $f(Q1) = 1khz$

   b) frequency at Q2 = f(Q2)

      $f(Q2) = (1 / 2^2) * f(CP)$
      $f(Q2) = (1/4) * 2khz$
      $f(Q2) = .5 khz$

   c) frequency at Q3 = f(Q3)

      $f(Q3) = (1 / 2^3) * f(CP)$
      $f(Q3) = (1/8) * 2khz$
      $f(Q3) = .25 khz$

In figure 2.5 at pulse 4 notice the propagation delay between the trailing edge of CP4 and b1. (A propagation delay is the time required for a logic circuit to respond to an input signal.). This propagation delay is caused by FF1 (the time required to transfer a signal from the inputs to the outputs of FF1). The same phenomena occurs between FF1 and FF2 (a2), and FF2 and FF3 (a3).

Even if the propagation delay of a flip-flop is very short, even in the order of nanoseconds (10E-9 seconds), problems emerge at a high frequency. The propagation delay of the flip-flops in a counter is additive. In other words the **total delay is the sum of the delays of each flip-flop. (Refer to figure 2.5. The** propagation delay is greater at CP8, a4). Therefore when a large number of flip-flops are used (to count to a large number), the propagation delay at the last flip-flop is large, and the counter may lag some input pulses. Therefore the final count will not represent the exact number of inputs. This problem can be solved by using synchronous counters (or parallel counters) which we will discuss later in this chapter.

Example

Generate a timing diagram (for 17 pulses) for the following counter. Show the propagation delay between the last flip-flop and the master clock applied to the counter inputs.

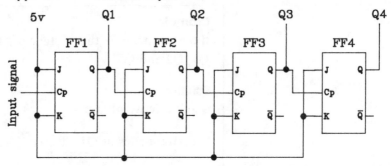

Figure 2.7. Modulo-16 binary counter

Solution

From figure 2.7, we see that FF1, FF2, FF3, and FF4 are in the toggle mode (J=K=1). Therefore the basic operation of the binary counter depicted in figure 2.7 resembles that of the binary counter shown in figure 2.4. The timing diagram of the modulo-16 binary counter is shown in figure 2.8. Note that the pulse trains of FF1, FF2, and FF3 are the same as those of the modulo-8 binary bounter shown in figure 2.5.

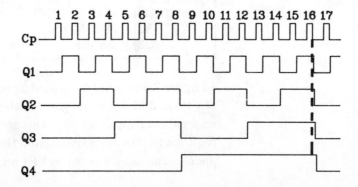

Figure 2.8. Timing diagram of a modulo-16 binary counter

**Practice
Problems**

Problem 2.1

What is the mode of operation of the flip-flops in a binary counter.

Solution

In a binary counter all the flip-flops are in the toggle mode.

Problem 2.2

Design a modulo 16 binary counter.

Solution

Use equation 2.1b to compute the number of flip-flops required to design a counter that counts to 16:

$$n = \ln(mod)/\ln(2) \qquad\qquad (2.1b)$$
$$n = \ln(16)/\ln(2)$$
$$n = 4 \text{ flip-flops are required}$$

The first step in the design of a binary counter is to make the flip-flops operate in the toggle mode by connecting the J and K inputs to a high signal (5 volts). Then connect the output Q of each flip-flop to the clock pulse (CP) input pin of the next flip-flop. The final counter resembles the following:

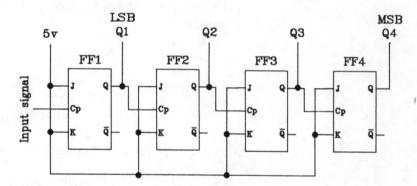

Figure 2.9. Modulo-16 binary counter

Problem 2.3

Define the period and the frequency of a signal.

Solution

A period is the time (in seconds) required for a signal to repeat itself. If a signal repeats itself every 5 milliseconds, then the period (T) is 5ms.

The frequency (F) is the inverse of the period (T). In other words, the frequency tells the number of occurrences of the signal in a time frame of 1 second. If a signal has a frequency of 500 hz it means that the signal repeats itself 500 times in 1 second.

Problem 2.4

If the frequency at Q2 is 2Khz, what is the frequency at the input of FF1* at Q1 and at Q3 for the following binary counter. (Hint use formula 2.2e.)

*f(CP)

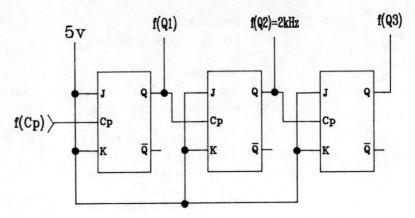

Figure 2.10. Modulo-8 binary counter

Solution

Let's first compute the frequency of the clock pulse (input frequency to be counted):

$$F(Qn) = (1 / 2^n) * F(CP)$$ (2.2e)

1) Input frequency:

$$F(CP) = F(Qn) * 2^n$$

with n=2 and F(Q2) = 2 khz we arrive at:

$$F(CP) = 2\ Khz * 2^2 = 8\ Khz$$

The input frequency is F(CP) = 8 khz.

2) Frequency at Q1

$$F(Q1) = (1/2^1) * 8\ khz$$
$$= 4\ khz$$

3) Frequency at Q3

$$F(Q3) = (1/2^3) * 8\ khz$$
$$= 1\ khz$$

BCD Counters

BCD counters (also called decade counters) are counters that have a maximum count of 10. The most commonly used BCD counters count in 8421 binary code. The binary count starts at 0000 and increments until it reaches 1001; then the counter is reset. The process repeats as shown in figure 2.11.

BCD	8	4	2	1	Decimal equivalent
	0	0	0	0	0
	0	0	0	1	1
	0	0	1	0	2
	0	0	1	1	3
	0	1	0	0	4
	0	1	0	1	5
	0	1	1	0	6
	0	1	1	1	7
	1	0	0	0	8
	1	0	0	1	9

Figure 2.11. 8421 BCD count

Figure 2.12 depicts an 8421 BCD counter. Note the changes in the configuration between the binary counter (figure 2.7) and the BCD counter (figure 2.12). In figure 2.12 the AND gate (gate a) controls the mode of operation of FF4, and the not Q output of FF4 controls the mode of operation of FF2. In the binary counter, all the flip-flops are in the toggle mode. Also the Cp input of FF4 comes directly from Q1 of FF1. These are the differences between binary and BCD counters. These changes reset the counter when the count reaches 1001 as we will see in the timing diagram of figure 2.13.

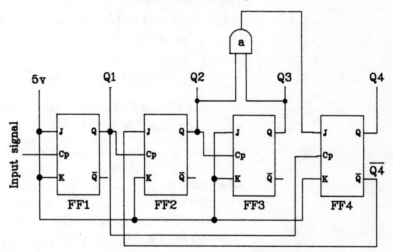

Figure 2.12. 8421 BCD counter

Note

The mode of operation of FF2 is controlled by the not Q output of FF4. Refer to figure 2.12 and note that the J input of FF2 is the same as not Q4 of FF4. The K input of FF2 is always high. Therefore if not Q4 is 1, J is also 1 (K is always 1), and FF2 is in the toggle mode (J=K=1). If not Q4 is 0, J is also 0 (K is always 1), and FF2 is in the reset mode (J=0), K=1). (Refer to table 2.1 for more details about these modes.)

The mode of operation of FF4 is controlled by the output of the AND gate (a). When FF2 and FF3 are set, Q2=Q3=1 and the output of the AND gate is 1. Therefore FF4 is in the toggle mode. When FF2, FF3, or both are reset, the output of the AND gate is 0. If this is the case FF4 is in the reset mode and Q4 is low.

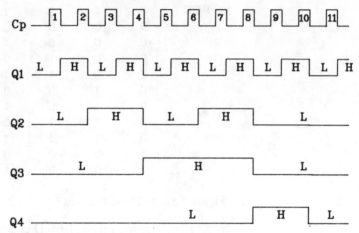

Figure 2.13. Timing diagram for a BCD counter

Timing Diagram Analysis of Figure 2.13

Before pulse 1

All the flip-flops are reset. Therefore Q4=Q3=Q2 =Q1=0. The BCD count is 0000.

After pulse 1

The high to low transition of CP1 sets FF1. Therefore Q1=1. FF2 is in the toggle mode (not Q4=1), but the low to high transition of Q1 won't affect FF2. (Recall that only a high to low transition affects a negative edge triggered clock.) Therefore FF2 remains reset and Q2=0. FF2 remains unchanged. Therefore FF3 remains reset and Q3=0.

Because FF2 and FF3 are reset, the output of gate a is 0 and FF4 is in the reset mode. Therefore Q4=0.

The count after CP1 is 0 0 0 1 (= Q4 Q3 Q2 Q1).

After pulse 2

The high to low transition of Q1 would set FF2 and Q2=1. The low to high transition of Q2 won't affect FF3. FF3 remains reset and Q3=0. Because FF3 is reset, the output from the AND gate a is 0. Therefore FF4 is in the reset mode. Thus FF4 remains reset and Q4=0.

The count after CP2 is 0 0 1 0.

After pulse 3

FF1 is set and Q1=1. FF2 doesn't change (low to high transition of Q1). Thus Q2=1. FF3 remains the same because FF2 didn't change. Therefore Q3=0. FF3 is reset. Therefore FF4 is in the reset mode and Q4=0.

The count after CP3 is 0 0 1 1.

After pulse 4

FF1 is reset and Q1=0. The high to low transition of Q1 resets FF2. Therefore Q2=0. The high to low transition of Q2 sets FF3. Therefore Q3=1. Since FF2 is reset, the output of the AND gate remains 0. Therefore FF4 is in the reset mode and Q4=0.

The count after CP4 is 0 1 0 0.

After pulse 5

FF1 is set and Q1=1. The low to high transition of Q1 won't affect FF2. Therefore FF2 remains unchanged and Q2=0. FF2 remains unchanged implying that FF3 won't change either. Then Q3=1. FF2 and FF3 are unchanged implying that FF4 remains unchanged. Q4=0.

The count after CP5 is 0 1 0 1.

After pulse 6

FF1 is reset and Q1=0. The high to low transition of Q1 toggles FF2. Therefore the output state of FF2 is complemented and FF2 is set. Q2=1. The low to high transition of Q2 won't affect FF3. Therefore Q3=1. Because FF2 and FF3 are both set, the output of the AND gate is 1 and FF4 is in the toggle mode. At the next high to low transition of Q1, FF4 will be complemented. For the moment FF4 remains unchanged and Q4=0.

The count after CP6 is 0 1 1 0.

After pulse 7

FF1 is set and Q1=1. The low to high transition of Q1 won't affect FF2. Therefore Q2=1. No change in FF2 implies no change in FF3. Therefore Q3=1. FF2 and FF3 are set; the output from gate a is 1 and FF4 is in the toggle mode. Note that the input to CP of FF4 comes from Q1. Q1 has gone from low to high. This transition of Q1 won't toggle FF4. Therefore FF4 remains unchanged and Q4=0.

The count after CP7 is 0 1 1 1.

After pulse 8

FF1 is reset and Q1=0. The high to low transition of Q1 toggles FF2. Therefore FF2 is reset and Q2=0. The high to low transition of Q2 toggles FF3. Therefore FF3 is reset and Q3=0. The high to low transition of Q1 toggles FF4. FF4 is set and Q4=1. Note that not Q4 is now 0; not Q4 is also the input J to FF2. Therefore FF2 will be in the reset mode until FF4 is reset.

The count after CP8 is 1 0 0 0.

After pulse 9

FF1 is set and Q1=1. The low to high transition of Q1 won't affect FF2. Therefore Q2=0. Because FF2 has not changed, FF3 won't either. Therefore Q3=0. Note that FF2 and FF3 are reset and that the output of the AND gate is

0, which causes FF4 to be in the reset mode. FF4 won't be reset by the low to high transition of Q1. Therefore FF4 remains set until a high to low transition of Q1 occurs. After CP9, Q4=1.

The count after CP9 is 1 0 0 1.

After pulse 10
FF1 is reset and Q1=0. FF2 remains reset because it is in the reset mode (input J=not Q4 is low). Therefore Q2=0. Because FF2 hasn't changed, FF3 will remain reset and Q3=0. The high to low transition of Q1 resets FF4. Therefore Q4=0. All the flip-flops are reset and the count starts from 0000 again.

The count after CP10 is 0 0 0 0.

After pulse 11
FF1 is set and Q1=1. The low to high transition of Q1 won't affect FF2. Therefore FF2 remains reset and Q2=0. Since FF4 was reset in CP10 (not Q4=1), FF2 is in the toggle mode again and a high to low transition of Q1 will now affect FF2. FF2 hasn't changed, thus FF3 also won't change. Therefore Q3=0. The output from the AND gate is 0 (FF2 and FF3 are reset). Therefore FF4 is in the reset mode (J=0 and K=1). The low to high transition of Q1 won't affect FF4. Therefore Q4=0.

The count is now 0 0 0 1.

Input Pulse #	Binary Count				Decimal Equivalent	
	Q4	Q3	Q2	Q1		
CP0	0	0	0	0	0	
CP1	0	0	0	1	1	
CP2	0	0	1	0	2	
CP3	0	0	1	1	3	
CP4	0	1	0	0	4	
CP5	0	1	0	1	5	
CP6	0	1	1	0	6	
CP7	0	1	1	1	7	
CP8	1	0	0	0	8	
CP9	1	0	0	1	9	
CP10	0	0	0	0	0	(RESET)
CP11	0	0	0	1	1	
.	.	.	.	.	.	
.		.	.	.	.	
.		.	.	.	.	
etc			etc		etc	

At CP10 the flip-flops are all reset and the count starts from 0000.

Since one BCD counter has a maximum count of 9, two BCD counters
cascaded as shown in figure 2.14 have a maximum count of 99. The first BCD
counter (counter 1) represents the LSD (least significant digit), and the second
counter (counter 2) represents the MSD (most significant digit).

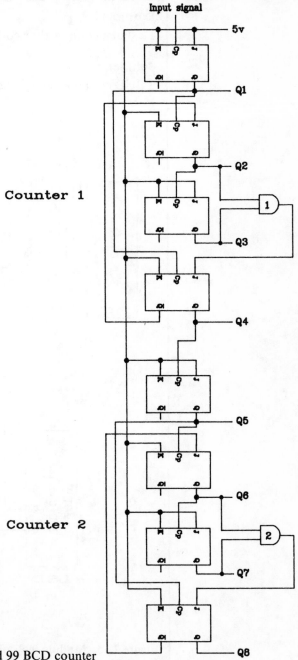

Figure 2.14. Mod 99 BCD counter

Operation

The operation of BCD counter 1 and BCD counter 2 (before they are connected) is identical to the BCD counter we examined earlier. Once these two counters are connected together to form one counter as in figure 2.14, the output of FF4 (Q4) triggers CP5. Therefore when the binary count in the first BCD counter reaches 1001 (maximum count for one BCD counter), the next clock pulse creates a high to low transition in Q4. This transition triggers FF5 and resets the flip-flops (FF1, FF2, FF3, and FF4) in the first BCD counter. Now the count is 0001 0000 (Q8 Q7 Q6 Q5 Q4 Q3 Q2 Q1 respectively) which corresponds to decimal 10. The count is now affecting the second counter.

Example
Design a counter that counts to 999.

Solution
3 BCD counters have to be cascaded to reach a count of 999. The counter is:

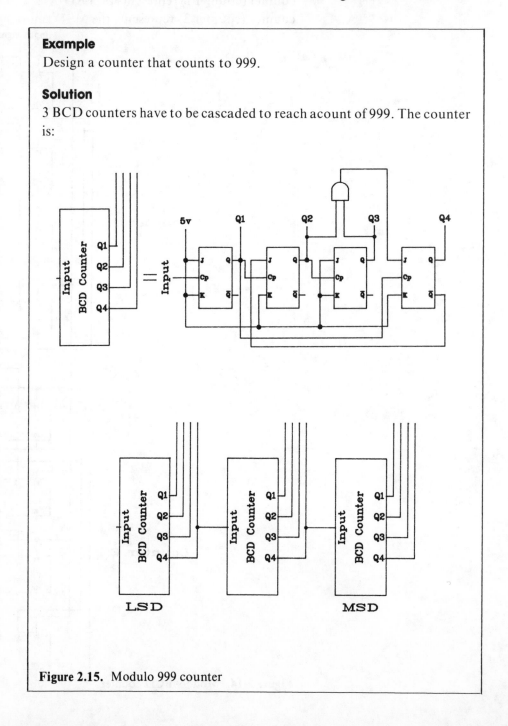

Figure 2.15. Modulo 999 counter

Frequency Consideration with BCD Counters

BCD counters are used when an input frequency needs to be divided by a multiple of 10.

Example

Suppose that we are designing a memory board for a CPU that runs at .5MHz. The memory chips that we are using are DRAM (Dynamic Random Access Memory). These DRAMs require a refresh signal every 2 milliseconds so they retain their data. Design a circuit that uses CPU frequency as input (.5MHz) and outputs the required frequency signal (.002 second) for the DRAM chips.

Solution

Let's first convert seconds into Hertz. Using formula 2.2c, we obtain:

f = 1/.002 = 500 Hz

The circuit characteristics are:

Output frequency is f(Out) = 500 Hz
Input frequency is f(In) = .5MHz = 500000 Hz

In other words the circuit should divide the input frequency signal by 1000 (1000 = 500000/500) to output the required signal (500 Hz) to refresh the DRAM chips. We need to cascade 3 BCD counters as follows:

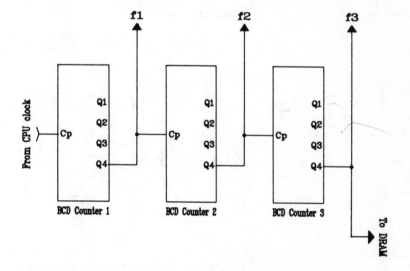

Figure 2.16. Interface circuitry

At f1 the frequency is 50000 Hz, at f2 the frequency is 5000 Hz, and at f3 the frequency is 500 Hz.

Problem 2.5

What is the difference between a BCD counter and a binary counter.

Solution

A BCD counter counts to a maximum of 10; in other words after the tenth pulse the counter is reset. A binary counter's maximum count is determined by the number of flip-flops contained in the counter. If a binary counter contains 2 flip-flops, the maximum count is 4, if the binary counter contains 3 flip-flops, the maximum count is 8 and so on.

Problem 2.6

Are all the flip-flops of a BCD counter in the toggle mode? Explain.

Solution

Only two flip-flops (FF1 and FF3) are in the toggle mode at all times (*see* figure 2.12). **The mode of FF2 is controlled by the not Q4 output of FF4. When** not Q4 is 1, FF2 is in the toggle mode. When not Q4 is 0, FF2 is in the reset mode. The mode of operation of FF4 is controlled by FF3 and FF2. FF4 is in the toggle mode if and only if both FF2 and FF3 are set. Otherwise FF4 is in the reset mode. This mode of operation of FF2 and FF4 is necessary to achieve a total reset at the tenth pulse.

Problem 2.7

Design a counter that counts to a maximum of 99.

Solution

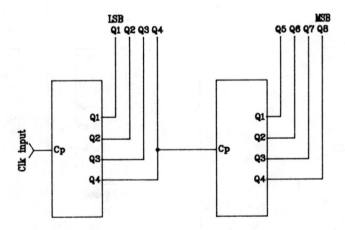

Figure 2.17. Binary counter that counts to decimal 99

Problem 2.8

How many BCD counters are needed to divide a signal frequency by 1 million?

Solution

6 BCD counters.

The output from the first BCD counter is f/10, where f is the frequency of the input signal. From the second BCD counter the output frequency is f/100. From the third BCD counter the output frequency is f/1000. From the fourth BCD counter the output frequency is f/10000. From the fifth BCD counter the output frequency is f/100000, and finally from the sixth BCD counter the output frequency is f/1000000.

Problem 2.9

Design a sequential circuit that outputs a frequency of 1kHz given an input frequency of 1MHz.

Solution

3 BCD counters are needed to achieve this operation.

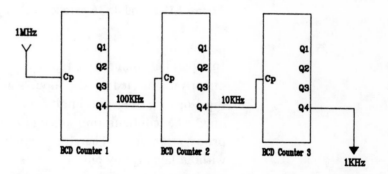

Figure 2.18. Frequency division with counters

Parallel Counters

Earlier, we discussed the time lagging problem of asynchronous binary counters. To refresh your memory, the propagation delay of the counter increases with a high frequency and the counter lags pulses. This propagation delay accumulation problem is solved by using sychronous binary counters (also called parallel counters).

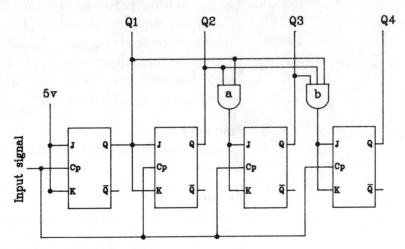

Figure 2.19. Modulo 16 parallel counter

In a parallel counter (*see* figure 2.19), each individual flip-flop's CP input is directly connected to the input signal to be counted. In other words all flip-flops are triggered simultaneously, making this operation a synchronous one. The configuration of figure 2.19 eliminates the additivity of the propagation delay. Therefore a count at a higher frequency can be obtained without lagging any pulses.

From figure 2.19 we can see that FF1 is in the toggle mode of operation at all times.

FF2's mode of operation is controlled by Q1 (the output of FF1). Therefore FF2 is in the toggle mode only if Q=1 (or FF1 is set). Otherwise if Q1=0, FF2 is in the hold mode.

FF3's mode of operation is controlled by the AND gate (gate a). Therefore FF3 is in the toggle mode of operation only if Q1 and Q2 (the inputs to the a gate) are high (or 1). In other words, FF2 and FF3 must be set. If either FF1 or FF2 or both are reset, the output from gate a is low (or 0) and FF3 is in the hold mode.

FF4 is in the toggle mode of operation only if FF1, FF2, and FF3 are set; otherwise FF4 is in the hold mode of operation.

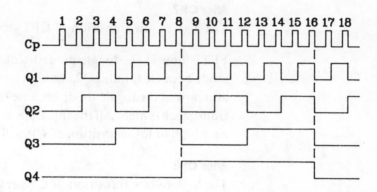

Figure 2.20. Timing diagram for a mod 16 parallel counter

Refer to figure 2.20 while following this step by step analysis of the operation of the mod-16 parallel counter of figure 2.19.

Before CP1
All the flip-flops are reset and Q1=Q2=Q3=Q4. The count is 0 0 0 0.

After CP1
The high to low transition of CP1 sets FF1. The count is 0 0 0 1.

Q1 is high. Therefore FF2 is in the toggle mode, and the next high to low transition of CP will set FF2.

After CP2
The high to low transition of CP2 resets FF1. The high to low transition of CP2 sets FF2. The count is 0 0 1 0.

After CP3
The high to low transition of CP3 sets FF1. The count is 0 0 1 1.

FF3 is now in the toggle mode because Q1 and Q2 are high, and the output from gate a is high (or 1). The next high to low transition of CP will set FF3.

After CP4
The high to low transition of CP4 resets FF1. The high to low transition of CP4 resets FF2. The high to low transition of CP4 sets FF3. The count is 0 1 0 0.

FF2 is now in the hold mode because Q1 is low. FF3 is now in the hold mode because Q1 and Q2 are low, and the output from gate a is low.

After CP5
The high to low transition of CP5 sets FF1. The count is 0 1 0 1. FF2 is now in the toggle mode because Q1 is high (or 1).

After CP6
The high to low transition of CP6 resets FF1. The high to low transition of CP6 sets FF2. The count is 0 1 1 0. FF2 is now in the hold mode because Q1 is low.

After CP7

The high to low transition of CP7 sets FF1. The count is 0 1 1 1.

FF2 is now in the toggle mode because Q1 is high. FF3 is now in the toggle mode because Q1 and Q2 are high and the output from gate a is high. FF4 is also now in the toggle mode because Q1, Q2, and Q3 are high, and the output from gate b is high. All the flip-flops are now in the toggle mode. Therefore the next high to low transition of CP will complement all the flip-flips.

After CP8

The high to low transition of CP8 resets FF1, FF2, FF3, and sets FF4. The count is 1 0 0 0.

Q1 is low and FF2 is in the hold mode. The output from gate a is low. FF3 is in the hold mode. The output from gate b is low. FF4 is in the hold mode.

After CP9

The high to low transition of CP9 sets FF1. The count is 1 0 0 1. Q1 is high and FF2 is in the toggle mode.

After CP10

The high to low transition of CP10 resets FF1, and sets FF2. The count is 1 0 1 0. Q1 is low and FF2 is in the hold mode.

After CP11

The high to low transition of CP11 sets FF1. The count is 1 0 1 1. Q1 is high and FF2 is in the toggle mode. Q1 and Q2 are high and FF3 is in the toggle mode.

After CP12

The high to low transition of CP12 resets FF1, FF2, and sets FF3. The count is 1 1 0 0. Q1 is low and FF2 is in the hold mode.

After CP13

The high to low transition of CP13 sets FF1. The count is 1 1 0 1. Q1 is high and FF2 is in the toggle mode.

After CP14

The high to low transition of CP14 resets FF1 and sets FF2. The count is 1 1 1 0. Q1 is low and FF2 is in the hold mode.

After CP15

The high to low transition of CP15 sets FF1. The count is 1 1 1 1.

Q1 is high and FF2 is in the toggle mode. The output from gate a is high; therefore FF3 is in the toggle mode. The output from gate b is high; thus FF4 is in the toggle mode. All of the flip-flops are set. Furthermore the flip-flops are all in the toggle mode. The next high to low transition of CP will reset all of the flip-flops.

After CP16

The high to low pulse of CP16 resets FF1, FF2, FF3 and FF4. The count is 0 0 0 0. All the flip-flops were reset. The count is recycled. Now the count will start again from 0 0 0 0.

After CP17

Same as "after CP1."

After CP18

Same as "after CP2."

In a parallel counter all of the flip-flop's CP inputs are tied to the input signal to be counted as follows:

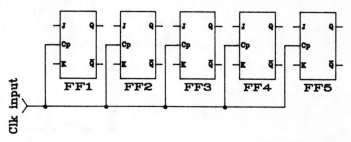

Figure 2.21. CP inputs

All the flip-flops in a parallel counter are in either the toggle (J=K=1) or hold (J=K=0) modes with the exception of the first flip-flop (FF1 or LSB flip-flop), which is in the toggle mode at all times. The mode of operation of a flip-flop (toggle or hold) is determined by the state of the preceding flip-flops. For example, the mode of operation of FF2 is detemined by the state of FF1. If FF1 is set then FF2 is in the toggle mode; if FF1 is reset then FF2 is in the hold mode.

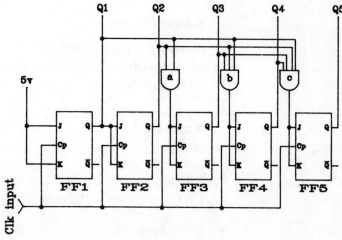

Figure 2.22. J-K inputs

FF3's mode of operation is determined by the state of both FF1 and FF2. If both FF1 and FF2 are set (Q1=Q2=1), the output of gate a is high and FF3 is in the toggle mode. Otherwise FF3 is in the hold mode of operation.

FF4's mode of operation is determined by the state of FF1, FF2, and FF3. If FF1, FF2, and FF3 are set (Q1=Q2=Q3=1), the output of gate b is high and FF4 is in the toggle mode. Otherwise FF4 is in the hold mode.

FF5's mode of operation is determined by the state of FF1, FF2, FF3 and FF4. If FF1, FF2, FF3, and FF4 are all set (Q1=Q2=Q3=Q4=1), the output of gate c is high and FF5 is in the toggle mode. Otherwise FF5 is in the hold mode of operation.

An AND gate is used to control the mode of operation of a flip-flop. If we're controlling FFn (where n is the flip-flop's number), the AND gate to be used should have (n−1) inputs. The input signals to the AND gate should originate at the Q output of the preceding flip-flops.

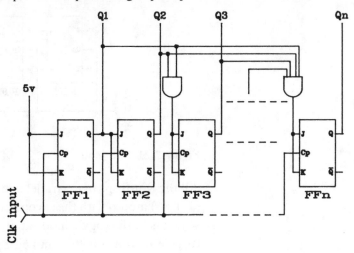

Practice Problems

Problem 2.10
Discuss the advantage of synchronous counters over asynchronous ones.

Solution
Because the propagation delay in an asynchronous counter is additive, the frequency of counting is limited. With a synchronous counter the propagation delay is not additive. Therefore the frequency of counting can be higher.

Problem 2.11
Design a mod 8 parallel counter.

Solution
Let's first compute the number of flip-flops needed:

Using equation 2.2

$$n = \ln(mod) / \ln(2)$$
$$= \ln(8) / \ln(2)$$
$$n = 3 \text{ flip-flops}$$

Now let's design the counter:

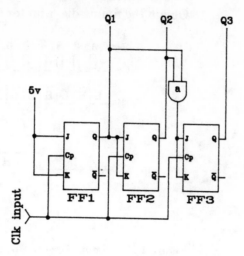

Figure 2.23. Modulo-8 parallel counter

Problem 2.12
Repeat 2.11 for a mod 128.

Solution
The number of flip-flops needed is:

$$n = \ln(mod) / \ln(2)$$
$$= \ln(128) / \ln(2)$$
$$= 7$$

A modulo 128 parallel counter follows:

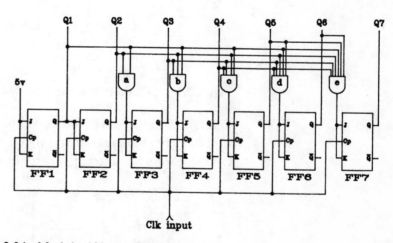

Figure 2.24. Modulo-128 parallel counter

Problem 2.13

Obtain the timing diagram for the counter of problem 2.11 for 17 pulses.

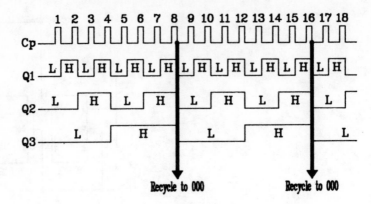

Figure 2.25. Timing diagram for a modulo-8 parallel counter

Other Binary Counters

Binary Down Counters

A down counter is the opposite of the up counters discussed previously. A 3 bit binary down counter starts at 1 1 1; each time an input signal is sensed, the counter is decremented by 1. This continues until the binary number stored by the counter becomes 0 0 0. The counter's flip-flops are then set so the count again starts at 1 1 1. The counter starts decrementing again (*see* figure 2.26).

Binary Count			Decimal
Q3	Q2	Q1	quivalent
1	1	1	7
1	1	0	6
1	0	1	5
1	0	0	4
0	1	1	3
0	1	0	2
0	0	1	1
0	0	0	0

Figure 2.26. 3 bit down counter

Figure 2.27 represents a 3 bit down counter. Note that the CP input signals of FF2 and FF3 eminate from the not Q output of FF1 and FF2 respectively. The CP input for up counters eminates from the Q outputs.

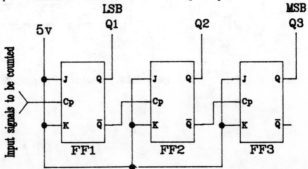

Figure 2.27. A 3 bit down counter

Figure 2.28 represents a timing diagram for a 3 bit down counter. Note that the first clock pulse sets all the flip-flops (Q1= Q2=Q3=1). The next pulse decrements the count by 1. The count becomes 1 1 0. Every input pulse causes the binary count stored by the counter to be decremented until the binary count reaches 0 0 0. When the count reaches 0 0 0, the counter's flip-flops are set so the count is again 1 1 1. This process is repeated as long as input pulses are applied to the counter.

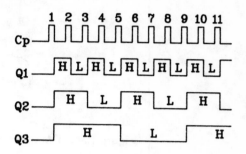

Figure 2.28. Timing diagram for a 3 bit down counter

Example
Develop a 2 bit binary down counter.

Solution
A 2 bit counter is shown in figure 2.29.

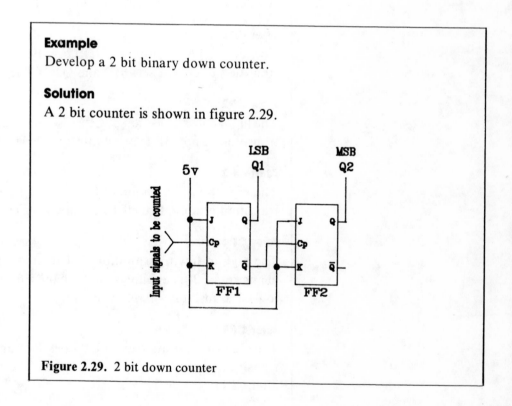

Figure 2.29. 2 bit down counter

Example

Obtain the timing diagram of a 2 bit down counter for 7 input pulses.

Solution

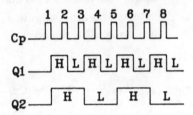

Figure 2.30. Timing diagram for 2 bit down counter

Before CP1

All the flip-flops are reset. The count is 0 0 (Q2 Q1).

After CP1

The high to low transition of CP1 sets FF1 and Q1=1. The high to low transition of not Q1 sets FF2 and Q2=1. The count is 1 1.

After CP2

The high to low transition of CP2 resets FF1 and Q1=0. The low to high transition of not Q1 does not affect FF2. The count is 1 0.

After CP3

The high to low transition of CP3 sets FF1 and Q1=1. The high to low transition of not Q1 resets FF2 and Q2=0. The count is 0 1.

After CP4

The high to low transition of CP4 resets FF1 and Q1=0. There is no change in FF2. The count is 0 0. FF1 and FF2 are reset. The next high to low CP transition will set the counter.

After CP5

The high to low transition of CP5 sets FF1 and Q1=1. The high to low transition of not Q1 sets FF2 and Q2=1. The count is 1 1. (The same as after CP1).

After CP6

The high to low transition of CP6 resets FF1 and Q1=0. There is no change for FF2; Q2=1. The count is 1 0. The counter's state after CP6 is the same as after CP2.

After CP7

Same as "after CP3."

Let's design a modulo-9 counter. A modulo-9 counter is a counter that resets itself after 9 pulses. In other words a mod-9 counter counts to a maximum of 8 (or binary 1 0 0 0) and then resets. The count sequence of a mod-9 counter should be as follows:

Q4	Q3	Q2	Q1	Decimal Equivalent
0	0	0	0	0
0	0	0	1	1
0	0	1	0	2
0	0	1	1	3
0	1	0	0	4
0	1	0	1	5
0	1	1	0	6
0	1	1	1	7
1	0	0	0	8
1	0	0	1	
1	0	1	0	
1	0	1	1	
1	1	0	0	
1	1	0	1	
1	1	1	0	
1	1	1	1	

The top of the table has the heading **Binary Count** spanning Q4 Q3 Q2 Q1.

Figure 2.31. Counting sequence of a mod-9 up counter

The counter should reset after 1000. In other words 1001 should cause the counter to reset (Q4=Q3= Q2=Q1=0). We need a combinational circuit that will reset FF1, FF2, FF3, and FF4 when outputs Q4 Q3 Q2 Q1 are 1 0 0 1.

Figure 2.32 represents a combinational circuit that must be added to a mod-16 counter so that the counter behaves as a mod-8 counter.

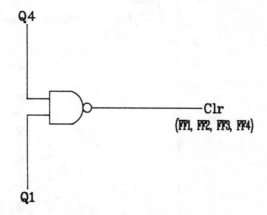

Figure 2.32. Interface circuit for a mod-9 counter

Note that the NAND gate has only two inputs which eminate from Q4 and Q1. These inputs (Q1 and Q4) to the NAND gate are sufficient because when *the first time* Q1 and Q4 are high (Q1=Q4=1), the counter is reset. The output from the NAND gate will go to the CLR input pin on all the flip-flops. Remember from chapter 1 that the asynchronous mode overrides the synchronous one. Remember also that CLR is low true. In other words a low signal (or 0) applied to the CLR will cause the flip-flop to reset (or clear). When Q4 and Q1 are high, the output of the NAND gate is low and FF1, FF2, FF3, and FF4 are reset. The counter is reset and the count starts again at 0 0 0 0.

The final mod-9 counter is shown in figure 2.33.

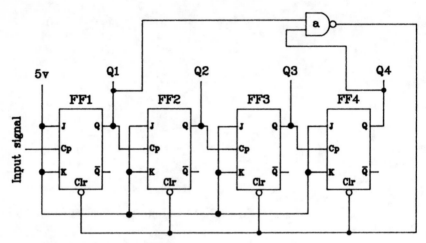

Figure 2.33. Modulo-9 up counter

Example

Design a mod-6 counter.

Solution

Let's first obtain the count sequence.

Binary Count			Decimal
Q3	Q2	Q1	Equivalent
0	0	0	0
0	00	1	1
0	1	0	2
0	1	1	3
1	0	0	4
1	0	1	5
1	1	0	
1	1	1	

The binary count immediately after the highest count of a modulo-6 counter is 1 1 0.

(cont. on page 97)

The inputs to the NAND gate should come from Q3 and Q2. Q2=Q3=1 should reset the circuit. The output should go to the CLR pin of FF1, FF2, and FF3.

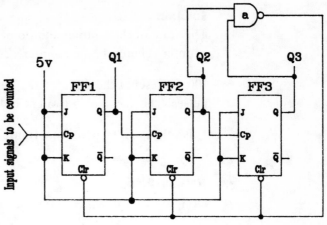

Figure 2.34. Modulo-6 up counter

Example

Obtain the timing diagram for 10 pulses using a modulo-6 counter.

Solution

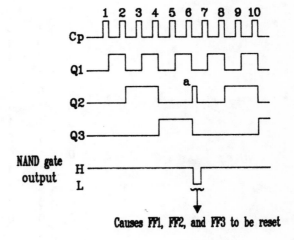

Figure 2.35. Timing diagram for a mod-6 counter

The a pulse in figure 2.35 causes the low pulse in the NAND gate. This low signal (output) from the NAND gate is fed back to the CLR input of FF1, FF2, and FF3. Therefore all the flip-flops are reset, and the count starts from 0 0 0. Note that the timing diagram of figure 2.35 does not include the delay of the flip-flops and the delay of the NAND gate.

Problem 2.14

A 3 bit down counter state is 1 0 0. What is the state of the counter after two pulses?

Solution

The state of the counter should be decremented by two because it is a down counter. Therefore the counter's count after two pulses is 100-010=010.

100 represents the state of the counter, and 010 represents the number of pulses. The decimal equivalent is 4-2=2.

This solution is evident from figures 2.26 or 2.28. After two pulses the state is 0 1 0.

Problem 2.15

Design a 4 bit down counter with its count sequence table.

Solution

The count sequence table of a 4 bit down counter follows:

Binary Count				Decimal
Q4	Q3	Q2	Q1	Equivalent
1	1	1	1	15
1	1	1	0	14
1	1	0	1	13
1	1	0	0	12
1	0	1	1	11
1	0	1	0	10
1	0	0	1	9
1	0	0	0	8
0	1	1	1	7
0	1	1	0	6
0	1	0	1	5
0	1	0	0	4
0	0	1	1	3
0	0	1	0	2
0	0	0	1	1
0	0	0	0	0

The design of a 4 bit binary down counter follows:

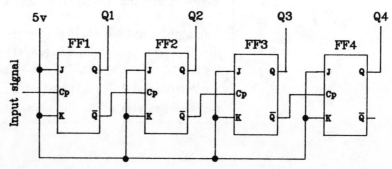

Figure 2.36. 4 bit down counter

Problem 2.16

Obtain the timing diagram for a 4 bit down counter. (Use 18 pulses.)

Solution

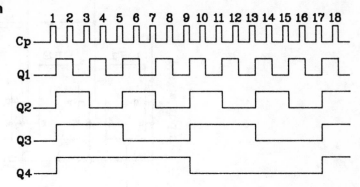

Figure 2.37. Timing diagram for a 4 bit down counter

Problem 2.17

Design a modulo-21 up counter. Give its count sequence table.

Solution

The count sequence for a mod-21 up counter follows:

Binary Count					Decimal Equivalent
Q5	Q4	Q3	Q2	Q1	
0	0	0	0	0	0
0	0	0	0	1	1
0	0	0	1	0	2
0	0	0	1	1	3
0	0	1	0	0	4
0	0	1	0	1	5
0	0	1	1	0	6
0	0	1	1	1	7
0	1	0	0	0	8
0	1	0	0	1	9
0	1	0	1	0	10
0	1	0	1	1	11
0	1	1	0	0	12
0	1	1	0	1	13
0	1	1	1	0	14
0	1	1	1	1	15
1	0	0	0	0	16
1	0	0	0	1	17
1	0	0	1	0	18
1	0	0	1	1	19
1	0	1	1	0	20
1	0	1	1	1	
1	1	0	0	0	
1	1	0	0	1	

...
...
etc

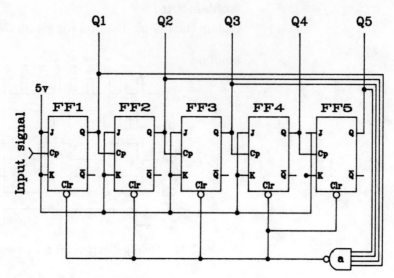

Figure 2.38. A mod-21 up counter

Problem 2.18

A CPU's clock runs at 4MHz. The CPU's communication board runs at 800kHz. Design a circuit that uses the CPU's clock signal to clock the communication board.

Solution

The circuit to be designed should divide the frequency of the CPU's clock by:

$$4MHz / 800kHz = 4E6 / 800E3 = 5$$

Therefore a mod-5 counter should be used.

The count sequence for a mod-5 counter follows:

Binary Count			Decimal
Q3	Q2	Q1	Equivalent
0	0	0	0
0	0	1	1
0	1	0	2
0	1	1	3
1	0	0	4
1	0	1	
1	1	0	
...			
etc			

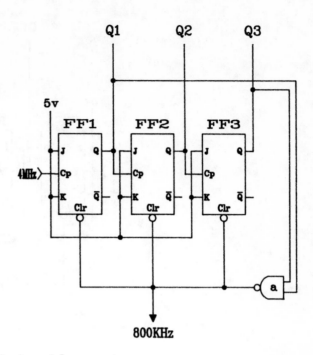

Figure 2.39. A mod-5 up counter

Experiment 4

Purpose

To build and demonstrate the operation of a mod-6 up counter.

Parts List

Quantity	Parts Description
1	Temporary board
1	5v power supply (or battery)
3	Resistors (200 ohms each)
3	Leds
1	Switch
1	Push button
1	7400 IC (two-input NAND gate)
2	7476 IC (dual J-K flip-flops)

Design Procedure

On page 97, we examined how a mod-6 up counter operates in theory. In this section we'll actually build a mod-6 up counter and then experiment with it.

Figure 2.39 represents a mod-6 up counter, and figure 2.41 represents the practical circuit to be built. Switch S1 of figure 2.41 will be used to input pulses into the counter. The number of pulses will be displayed by the leds.

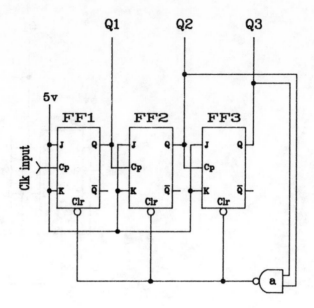

Figure 2.40. Mod-up counter

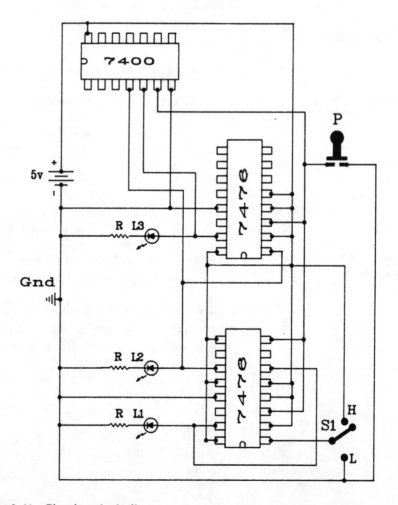

Figure 2.41. Circuit to be built

Operating Procedure

Wire the circuit of figure 2.41(refer to the earlier experiments for step by step wiring procedures).

1 Place switch S1 in the high position; then connect the battery. Examine the leds; if one or more leds are on, push P to reset the three flip-flops. Once the flip-flops are reset, all leds should be off. If not, disconnect the battery and check your circuit connections.

2 Once step 1 has been successfully completed, move switch S1 from the high to the low position. This transition (H to L) sets FF1. L1 is turned on. The counter is storing the binary value 0 0 1. L3=off; L2=off; L1=on.

Switch S1 back to the high position. The low to high transition doesn't affect the counter. Therefore the state of the flip-flop remains unchanged. Each time we switch from low to high, the counter won't be affected.

Continue the testing (second transition). One transition is considered to be from high to low to high.

Transition # (H-L-H)	Leds			Binary Count			Decimal Equivalent
	L3	L2	L1	Q3	Q2	Q1	
"Reset"	off	off	off	0	0	0	0
First	off	off	on	0	0	1	1
Second	off	on	off	0	1	0	2
Third	off	on	on	0	1	1	3
Fourth	on	off	off	1	0	0	4
Fifth	on	off	on	1	0	1	5
Sixth	off	off	off	0	0	0	0
Seventh	off	off	on	0	0	1	1
Eighth	off	on	off	0	1	0	2
Ninth	off	on	on	0	1	1	3
Tenth	on	off	off	1	0	0	4
Eleventh	on	off	on	1	0	1	5
Twelfth	off	off	off	0	0	0	0
Thirteenth	off	off	on	0	0	1	1
Fourteenth	off	on	off	0	1	0	2
Fifteenth	off	on	on	0	1	1	3
.	.	.	.	.	.	.	.
.	.	.	.	.	.	.	.
etc		etc			etc		etc

Problems

Problem 2.1

Define a counter.

Problem 2.2

Design a modulo 32 binary counter.

Problem 2.3

If the frequency at the Cp input is 4Mhz, what is the frequency at Q1, Q2, and Q3 of the following binary counter:

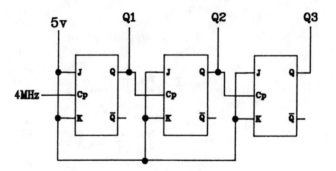

Problem 2.4

Design a counter that counts from decimal 0 to decimal 9999. (Use BCD counters.)

Problem 2.5

Design a sequential circuit (BCD counter) that outputs a frequency of 1 Hz given an input frequency of 100 Hz.

Problem 2.6

Design a modulo-16 parallel counter.

Problem 2.7

Design a 5 bit binary down counter with its count sequence table.

Problem 2.8

Obtain the timing diagram for a 3 bit down counter for 10 pulses.

Problem 2.9

Design a modulo-14 binary counter.

Problem 2.10

Design a modulo-14 binary down counter.

3

Shift Registers

Introduction

Shift registers are the most often used sequential circuits. Shift registers are vital to digital computers because of their ability to perform many logic operations. Shift registers can operate as a memory device (to store data); they can shift data (to the right or left); they can convert data (from serial to parallel, or from parallel to serial); they can do arithmetic operations (multiplication and division;)...etc.

Shift Register Basics

Shift registers are made up of flip-flops; S-R, D or J-K flip-flops can be used in the design of shift registers. A 4-bit shift register is shown in figure 3.1. Note that the shift register of figure 3.1 is implemented with D flip-flops.

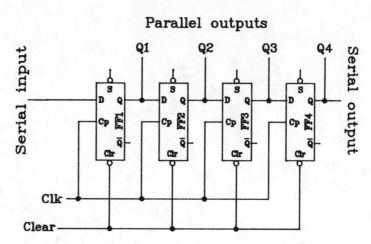

Figure 3.1. 4-bit right shift register

Figure 3.1 represents a 4-bit shift register. Note that the input to this shift register is serial (one bit is loaded at a time), and the output from this shift register can be either parallel (four bits are displayed from Q1 Q2 Q3 Q4) or serial (all the bits are displayed from Q4.)

Serial Input Defined

With serial input only one input line exists. Therefore the binary bits (data) are loaded one at a time. In other words, to load a single bit into the shift register one clock pulse is required; the second bit requires a second clock pulse to be loaded; ...etc. Therefore to load four bits into the shift register of figure 3.1 four clock pulses are required.

Serial Output Defined

Serial output uses one output line to output all the binary bits stored in the shift register. Each single binary bit stored in the shift register requires one clock pulse to be displayed. If the four binary bits stored into the shift register are output serially, four clock pulses must be applied. The binary bits emerge from the serial output one after the other. The serial output line is also the last flip-flop's output (Q4). In figure 3.1, by applying four clock pulses, the binary bits stored in the shift register will be shifted to the right and displayed at Q4.

Parallel Output Defined

Q1, Q2, Q3, and Q4 represent the parallel outputs of the shift register. A parallel output contains as many output lines as there are flip-flops in the shift register. In figure 3.1, we have four flip-flops; therefore four output lines are required. One clock pulse is needed to display the contents of the shift register (recall that four clock pulses are required for a serial output).

The advantage of parallel output over serial output is the time savings (one clock pulse for the parallel and 4 clock pulses for the serial). The advantage of serial output over parallel output is that fewer lines are required for the output (one line for a serial output and four lines for a parallel output).

Basics of Shift Register Operation

The shift register of figure 3.1 is composed of four D flip-flops. The output of each flip-flop is tied to the D input of the next immediate D flip-flop. This configuration allows the transfer of data from one flip-flop to the other.

As an example let's load the decimal 10 into the shift register of figure 3.1. The binary equivalent of decimal 10 is 1010. The loading process of the binary number 1010 into the shift register of figure 3.1 is shown in figure 3.2.

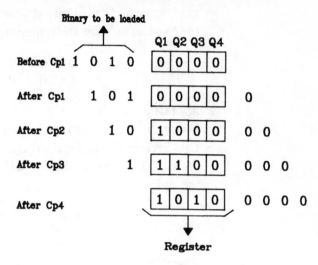

Figure 3.2. 4-bit right shift register operation

Recall that the input to the shift register is serial. Therefore each binary bit requires one clock pulse. To load the binary 1010 four clock pulses are required.

Before CP1

The shift register is cleared, and Q1=Q2=Q3=Q4=0. To clear the shift register, a low signal (or ground) should be applied to the *clr* input in figure 3.1. Once the shift register is cleared, the clr input should be high. (A 5 volt signal is to be applied to the clr input during normal operation of the shift register.) Throughout this example, parallel output is used.

Q1	Q2	Q3	Q4
0	0	0	0

After CP1

After the first clock pulse, binary 0 is loaded into the shift register. The serial input line is kept low to load bit 0 into the shift register. The contents of the shift register were shifted one space to the right to make room for the loaded bit. The binary number 0 stored in the rightmost D flip-flop (Q4) is lost, and the 0 stored in FF3 is shifted to FF4.

Q1	Q2	Q3	Q4
0	0	0	0

After CP2

To load binary bit 1 into the shift register, the serial input line should be high when CP2 arrives. After the second clock pulse, binary 1 is loaded. Note that the binary bit loaded into the shift register after CP1 was shifted to the right (one space) as were the remaining bits stored in the shift register.

Q1	Q2	Q3	Q4
1	0	0	0

After CP3

Binary bit 0 is to be loaded. Therefore the serial input line must be low when CP3 arrives. After the third clock pulse, binary bit 0 is loaded into the shift register. The binary bits stored in the shift register before CP3 were shifted to the right.

Q1	Q2	Q3	Q4
0	1	0	0

After CP4

To load the binary bit 1, the serial input line must be high when CP4 arrives. After the fourth clock pulse, binary 1 is loaded into the shift register. All the bits stored in the shift register prior to CP4 were shifted to the right to make room for the last binary bit.

Q1	Q2	Q3	Q4
1	0	1	0

The decimal number 10 is now loaded into the shift register. Note that with each clock pulse, the data was shifted one space to the right, and a binary bit was loaded into the shift register. If no data is available when the clock pulse signal arrives, the contents of the shift register will be shifted to the right and binary bit 0 will be loaded.

If we applied one more clock pulse after CP4, the binary number stored would become:

Q1	Q2	Q3	Q4
0	1	0	1

After the fifth clock pulse the binary number stored in the shift register would be 0101. In the decimal number system, the binary 0101 is equivalent to 5. Note that by applying an extra clock pulse to the shift register, the value stored in it was divided by 2.

First we used the shift register of figure 3.1 to load and store the decimal number 10. Then we used the shift register to divide the number stored in the shift register by 2.

Shift registers can undertake other logic operations as well. In the next sections shift register design and operation will be discussed in more detail.

Controlled Shift Registers

Shift Register Operation

A shift register is made up of a certain number of flip-flops. The number of flip-flops used determines the storage or shifting of a register. A shift register made of 4 flip-flops has a maximum storage of 4 bits. If we want to store a word, an 8 bit shift register is needed and 8 flip-flops are required. Flip-flops in a shift register are cascaded so that every binary bit can be shifted from one flip-flop to the other. J-K flip-flops are widely used in the design of shift registers because of their multiple modes of operation (set, reset, toggle, and hold).

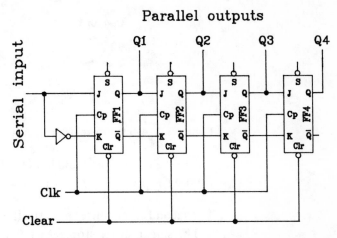

Figure 3.3. 4-bit shift register with J-K flip-flops

Figure 3.3 represents a 4-bit shift register (shift right). Four J-K flip-flops are used. Note that the outputs Q1 and not Q1 of the first flip-flop are respectively the J and K inputs of the second flip-flop. The outputs Q2 and not Q2 of the second flip-flop are respectively the J and K inputs of the third flip-flop,..etc. This configuration allows the transfer (or shift) of data from one flip-flop to the other.

Q is always the complement of not Q, so if Q=0 not Q1=1 and vice versa. Therefore J and K also have to be complementary because J=Q and K=not Q. Thus all the J-K flip-flops in a shift register operate in the set (when J=1 and K=0), or reset (J=0 and K=1) modes.

Data is loaded into a shift register synchronously with the clock. Every clock pulse causes the contents of the shift register to be shifted one space to the right (or left), and at the same time a new bit is loaded into the shift register.

Let's assume that the shift register is cleared before the loading process begins. Therefore Q1=Q2=Q3=Q4=0. Now suppose we want to load the binary number 0101 into the shift register of figure 8.3. Four clock pulses are required. Figure 3.4 represents the loading process of 0 1 0 1 into the shift register.

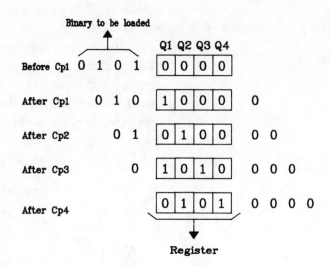

Figure 3.4. 4-bit shift register operation

The first clock pulse would load the first bit (LSB) or 1. The shift register stores 1 0 0 0 (respectively Q1 Q2 Q3 Q4) after the first clock pulse.

The second clock pulse shifts the first bit 1 to the right, and loads the second bit 0. The shift register stores 0 1 0 0 after the second clock pulse.

The third clock pulse shifts the contents of the shift register by one space to the right and loads the third bit 1. The shift register stores 1 0 1 0 after the third clock pulse.

The fourth clock pulse shifts the contents of the register by one space to the right and loads the fourth bit 0. After the fourth clock pulse, the shift register stores 0 1 0 1, completing the loading process.

If we apply a fifth clock pulse, the contents of the shift register would be shifted to the right by one space. Therefore the rightmost bit 1 (or LSB) of the shift register would be lost. The contents of the shift register would be 0 0 1 0.

The waveforms for loading the binary number 0101 into a 4-bit shift register appears in figure 3.5.

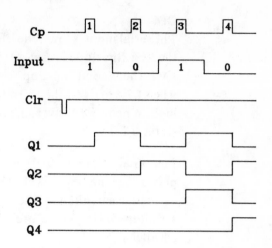

Figure 3.5. Timing diagram for loading 0101 into a 4-bit shift register

Figure 3.5 represents the timing diagram for a 4-bit shift register. Let's analyze these waveforms:

Before CP1

The CLR input is low. Therefore all the J-K flip-flops are reset and Q1, Q2, Q3, and Q4 are low. The shift register is said to be cleared (no data is stored) and Q1=Q2=Q3=Q4=0. Note that the clr input line went low for a short moment only; after that a high signal was applied to the clr input to disable it so that the shift register could operate properly.

CP1

The input signal is high, the first flip-flop is set (J=1 and K=0). Q1 is now high (or 1), and not Q is low. Binary bit 1 is loaded. The shift register is storing 1 0 0 0 (respectively Q1 Q2 Q3 and Q4).

CP2

The serial input is low (or 0). Therefore the first flip-flop is reset (J=0, K=1) and Q1=0. Before CP2 arrives, Q1=1 and not Q1=0. When the second clock pulse arrives, the second flip-flop is set, and Q2=1 and not Q2=0. The shift register stores 0 1 0 0 (respectively Q1 Q2 Q3 and Q4).

CP3

The serial input is high (or 1), the first flip-flop is set (J=1 and K=0), and Q1=1. Before CP3, Q1 was low. At the arrival of CP3 the second flip-flop is reset. Q2=0. Before CP3, Q2 was high. At the arrival of CP3 the third flip-flop is set and Q3=1. The shift register stores 1 0 1 0.

CP4

The serial input is low. Therefore the first flip-flop inputs are J=0 and K=1 and the first flip-flop is reset. Q1=0. Before CP4, Q1 was high. At the arrival of CP4, the second flip-flop is set. Q2=1. Before CP4, Q2 was low. Therefore when CP4 arrives, the third flip-flop is reset and Q3=0. Before CP4, Q3 was high. When CP4 arrives, the fourth flip-flop is set and Q4=1. The shift register stores 0 1 0 1.

The binary number 0101 was loaded into the shift register after four clock pulses. The input to this shift register is serial, because a clock pulse is required to load each bit. On the contrary, the outputs of this register are seen simultaneously. Therefore, the output of the shift register of figure 3.3 is parallel.

The shift register of figure 3.3 can be used as a data converter. In other words if serial data is to be converted into parallel data, the shift register of figure 3.3 could be used. In the next section, serial to serial, parallel to serial, and finally parallel to parallel shift registers are discussed.

Example
Design a 3-bit shift right register using J-K flip-flops.

Solution

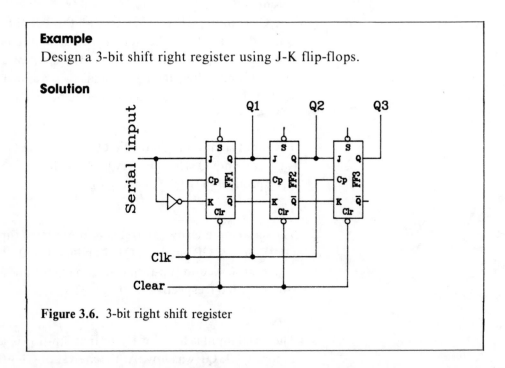

Figure 3.6. 3-bit right shift register

Compare the shift register of figure 3.6 and the shift register of figure 3.1. Both shift registers are 3-bit and their operation is the same.

Recall that the D flip-flip operation is controlled by the input D. If D is low, the D flip-flop is reset and Q=0; if the D input is high, the D flip-flop is set and Q=1. Even if two different types of flip-flops are used in figure 3.1 and 3.6, the operation of the shift registers remains the same because the J-K flip-flops of figure 3.6 operate as D flip-flops. The operation of a J-K flip-flop as a D flip-flop is caused by the NOT gate (*see* figure 3.6) in the first flip-flop, and by making the J and K inputs complementary in the other flip-flops. By making J and K complementary (J=Q and K=not Q), only two input combinations are possible J=0, K=1 (reset) and J=1, K=0 (set). Only two modes are available to the J-K flip-flops in figure 3.6.

Even if the circuit of figure 3.6 is more complex than the one depicted in figure 3.1, J-K flip-flops are more often used in the design of shift registers because of their ability to offer other modes of operation (toggle and hold modes) that could be used by the designer in some special cases.

Example

Describe pictorially the steps that occur when the binary number 110 is loaded into the shift register of figure 3.6.

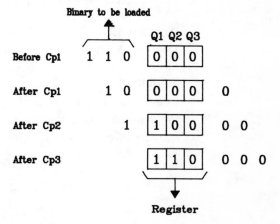

Figure 3.7. Loading binary 110 into a 3-bit shift register

Practice Problems

Problem 3.1

How many flip-flops are required to design a shift register capable of storing a word?

Answer

A word contains 8 bits. Each flip-flip can store 1 bit. Therefore 8 flip-flops are required to design a shift register that can store a word.

Problem 3.2

How many clock pulses are required to load a word in an 8-bit shift register? (The shift register's input is serial.)

Answer

In serial operation (input or output), one clock pulse is required to transfer one bit. Therefore a word requires 8 clock pulses to be loaded.

Problem 3.3

Design a 5-bit shift register using D flip-flops.

Solution

Five D flip-flops are required. Note that the output of each D flip-flop is tied to the input of the next immediate flip-flop. This configuration will allow the transfer of data from one flip-flop to the other.

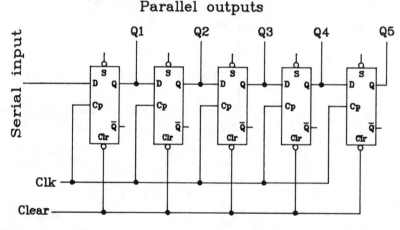

Figure 3.8. 5-bit shift register

Problem 3.4

Describe step by step the process for loading 11001 into a 5-bit shift register.

Solution

Before CP1

The shift register is cleared and $Q1=Q2=Q3=Q4=Q5=0$.

After CP1

Binary bit 1 is loaded into the shift register. The shift register outputs are:

1 0 0 0 0 (Q1 Q2 Q3 Q4 Q5).

After CP2

Binary bit 0 is loaded into the shift register. The shift register outputs are:

0 1 0 0 0 (Q1 Q2 Q3 Q4 Q5).

After CP3

Binary bit 0 is loaded into the shift register. The shift register outputs are:

0 0 1 0 0 (Q1 Q2 Q3 Q4 Q5).

After CP4

Binary bit 1 is loaded into the shift register. The shift register outputs are:

1 0 0 1 0 (Q1 Q2 Q3 Q4 Q5).

After CP5

Binary bit 1 is loaded into the shift register. The shift register outputs are:

1 1 0 0 1 (Q1 Q2 Q3 Q4 Q5).

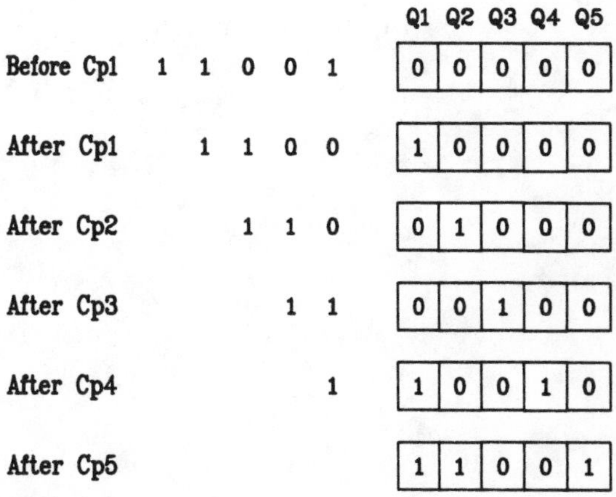

						Q1	Q2	Q3	Q4	Q5
Before Cp1	1	1	0	0	1	0	0	0	0	0
After Cp1		1	1	0	0	1	0	0	0	0
After Cp2			1	1	0	0	1	0	0	0
After Cp3				1	1	0	0	1	0	0
After Cp4					1	1	0	0	1	0
After Cp5						1	1	0	0	1

Figure 3.9. Operation of a 5-bit shift register

Problem 3.5

What binary bits are stored in the shift register of problem 3.4 after 8 clock pulses.

Solution

During the first 5 clock pulses, the binary bits 11001 are loaded (*see* problem 3.4), and the shift register resembles the following:

After CP5: 1 1 0 0 1

After loading 11001 into the shift register, the input line becomes low. If additional clock pulses are applied, 0's would be loaded.

When the sixth clock pulse is applied, the binary bit 0 would be loaded, and the shift register would resemble the following:

After CP6: 0 1 1 0 0

The rightmost binary bit (1) of the shift register in CP5 is lost. A seventh clock pulse would load another binary bit 0 into the shift register. The shift register would resemble the following:

After CP7:　0　0　1　1　0

If an eighth clock pulse is applied, the contents of the shift register would be shifted to the right, and a binary bit 0 would be loaded. Therefore the shift register would resemble the following:

After CP8:　0　0　0　1　1

Problem 3.6

Design a shift register that can store a word. (Use J-K flip-flops.)

Solution

Since a word contains 8 bits, eight J-K flip-flops are required. Note that each of the Q and not Q outputs are tied to the J and K inputs of the flip-flop to the immediate right. This configuration assures a transfer of data from one flip-flop to the other.

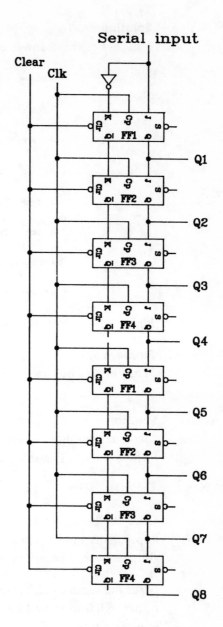

Figure 3.10. 8-bit shift register

Problem 3.7
Detail the steps needed to load 11000 1011 into the shift register of problem 3.6.

Solution

		Q1	Q2	Q3	Q4	Q5	Q6	Q7	Q8
Before Cp1	1 1 0 0 1 0 1 1	0	0	0	0	0	0	0	0
After Cp1	1 1 0 0 1 0 1	1	0	0	0	0	0	0	0
After Cp2	1 1 0 0 1 0	1	1	0	0	0	0	0	0
After Cp3	1 1 0 0 1	0	1	1	0	0	0	0	0
After Cp4	1 1 0 0	1	0	1	1	0	0	0	0
After Cp5	1 1 0	0	1	0	1	1	0	0	0
After Cp6	1 1	0	0	1	0	1	1	0	0
After Cp7	1	1	0	0	1	0	1	1	0
After Cp8		1	1	0	0	1	0	1	1

Figure 3.11. Steps for loading 1100 1011 into a 8-bit shift register

Problem 3.8

What binary bits are stored in the shift register of problem 3.7 after 10 clock pulses.

Solution

The binary number 11001011 was loaded into the shift register during the first 8 clock pulses. The shift register resembles the following:

After CP8:   1 1 0 0 1 0 1 1

Once the binary word was loaded, the input line becomes low (no signal is applied any more). Therefore if any additional clock pulses are applied to the shift register, a binary bit 0 will be loaded.

After the ninth clock pulse, the contents of the shift register are shifted to the right (LSB is lost), and a binary 0 is loaded into the shift register. Now the shift register outputs are:

After CP9:   0 1 1 0 0 1 0 1

If a tenth clock pulse is applied, the shift register's contents would be shifted to the right, and a binary 0 loaded. After the tenth clock pulse, the shift register's contents are:

After CP10:   0 0 1 1 0 0 1 0

Other Shift Registers

In the preceding section, we were exposed to shift registers that convert serial data to parallel. These are known as *serial in parallel out* shift registers. In this section we're going to learn to design and operate parallel-load shift registers. We will also study shift register applications such as their use in arithmetic operations.

Parallel-Load Shift Registers

If for example we wanted to load 8 bits of data into 8-bit *serial in prallel out* shift register; 8 clock pulses would be required. To load 2 words into a 16 bit shift register, 16 clock pulses are required. In other words, each bit to be loaded requires one clock pulse. When a large number of bits are to be loaded, a significant time is required for loading. This elapsed period of time can be eliminated by using a *parallel-load shift register*.

A parallel-load shift register is shown in figure 3.12. Note that the parallel inputs are asynchronous. In other words no clock pulses are required to input data into a parallel-load shift register. Furthermore the data can be loaded into a parallel-load shift register as a group of bits instead of one bit after the other (explanation follows):

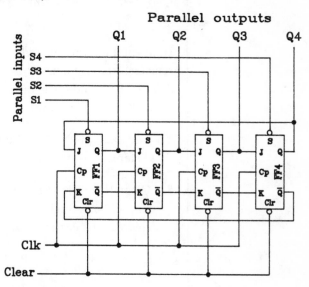

Figure 3.12. Parallel-load shift register

The contents of the shift register can be seen from Q1, Q2, Q3 and Q4. The input lines are tied to the S (or set) input pins of the flip-flops (S1, S2, S3, and S4) as shown in figure 3.12.

Note the feedback lines between the rightmost flip-flop or FF4 (the flip-flop that stores LSB) and the leftmost flip-flop or FF1 (the flip-flop that stores the MSB). Output Q4 is tied to the J input of the first flip-flop, and the not Q4 output is tied to the K input of the first flip-flop. This configuration allows the *circulation* of data through the shift register. With each clock pulse, the contents of the shift register will be shifted to the right, then back to the first flip-flop and so on. The data shifts from one flip-flop to the next never leaving the shift register. In other words, the data stored won't be lost when clock pulses are applied.

If new data is to be loaded, the shift register should be cleared first. To clear the shift register, a low signal should be applied to the clr line. Once the shift register is cleared, the parallel input lines (S1, S2, S3, and S4) are used to load the data into the shift register. The input lines are asynchronous and do not require a clock pulse.

Example

Let's load binary 1 0 1 0 into the shift register of figure 3.12.

To load binary 1 0 1 0 into the parallel-load shift register, we'll be using the asychronous inputs: S (set) and CLR (clear). Recall that the asynchronous inputs (S and CLR) operate independently of the clock. In other words the asynchronous inputs override all other inputs. Both asynchronous inputs (S and CLR) are low true. Low true means that a low signal activates (or enables) the input line. For example, when a low signal (or ground) is applied to the CLR input, the flip-flop is cleared and Q=0. When a low signal is applied to the S input, the flip-flop is set and Q=1.

In normal operations (when we're not loading data or clearing the shift register), the asynchronous inputs (CLR and S) should be disabled. To disable the S and CLR inputs, a high signal (or 5 volt) should be applied to the S and CLR inputs.

To load the binary 1 0 1 0 into the parallel-load shift register we follow these steps:

1   Clear the shift register first. To clear the shift register the clr line should be brought low for a short moment. Once the shift register is cleared, the contents of the shift register are:

Q1	Q2	Q3	Q4
0	0	0	0

2   Load binary 1010 into the shift register. In other words, FF1 and FF3 should be set (Q1=Q3=1). FF2 and FF4 should remain reset (Q2=Q4=0). To set FF1 and FF3, a low signal should be applied temporarily to the S inputs of FF1 and FF3 (respectively S1 and S3). The shift register's contents become:

Q1	Q2	Q3	Q4
1	0	1	0

In step 2, the loading of 1010 is completed. You may now wonder, "What happens when the first clock pulse arrives?" When the first clock arrives, the contents of FF1 will be shifted to FF2. The contents of FF2 will be transferred to FF3. The contents of FF3 will be transferred to FF4. And finally the contents of FF4 will be transferred to FF1.

The data *recirculates* through the shift register when clock pulses are applied. The data is never lost. To load new data into the shift register, we have to clear the shift register first, then load the binary bits. The following table shows the shift of data in the 4-bit parallel-load shift register in response to the first four clock pulses.

(cont. on page 125)

	Q1	Q2	Q3	Q4
Before CP1	1	0	1	0
After CP1	0	1	0	1
After CP2	1	0	1	0
After CP3	0	1	0	1
After CP4	1	0	1	0

Example

Design a 3-bit parallel-load shift register, load binary 110, and obtain the timing diagram for four clock pulses.

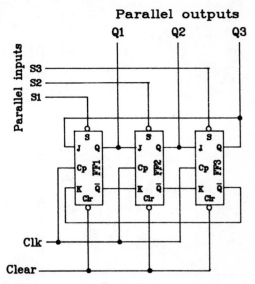

Figure 3.13. 3-bit parallel-load shift register

1   To load the binary 110 into the shift register, we should first clear the shift register. The input clr should go low for a moment. After the shift register is cleared we have:

Q1	Q2	Q3
0	0	0

2   To load the binary bits 110 into the parallel-load shift register, we have to apply a low signal to S1 and S2. The shift register's contents become

Q1	Q2	Q3
1	1	0

The data recirculates in the shift register during the first four clock pulses as follows:

	Q1	Q2	Q3
Before CP1	1	1	0
After CP1	0	1	1
After CP2	1	0	1
After CP3	1	1	0
After CP4	0	1	1

If we compare the values for Q1, Q2, and Q3 before CP1, and after CP3, we see that they are the same. Therefore three clock pulses were required to recirculate binary 110 to its original position. In general we need as many clock pulses as there are flip-flops in the shift register to recirculate the data to its original position. For a 4-bit parallel-load shift register, four clock pulses are required to recirculate the data to its original position. In general we need *n* clock pulses to recirculate the data to its original position in an n-bit parallel-load shift register.

The timing diagram for this operation is shown in figure 3.14.

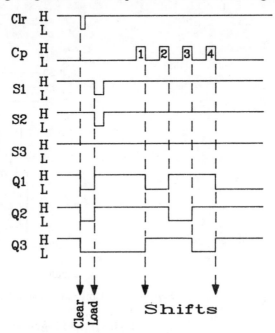

Figure 3.14. Waveforms for a 3-bit parallel-load shift register

Analysis of Figure 3.14

Clearing the Shift Register

Before CP1
(*See* CLEAR in figure 3.14. We apply a low signal to the CRL line (short moment). (*See* figures 3.13 and 3.14.)

The shift register's outputs become:

Q1	Q2	Q3
0	0	0

Loading the Binary Number

(*See* LOAD in figure 3.14). After clearing the shift register we load binary 110 into the shift register. We apply a low signal to S1 and S2 (short moment). (*See* figures 3.13 and 3.14.)

The shift register's outputs become:

Q1	Q2	Q3
1	1	0

Status of Flip-Flops

Before CP1

FF1: set; Q1=1, not Q1=0. (Notice that Q1 and not Q1 are also the J and K inputs to FF2.)

FF2: set; Q2=1, not Q2=0. (Notice that Q2 and not Q2 are also the J and K inputs to FF3.)

FF3: reset; Q3=0, not Q3=1. (Notice that Q3 and not Q3 are also the J and K inputs to FF1.)

After CP1

We now can predict what will happen when the first clock pulse arrives:

FF1: before CP1, J=0 and K=1. Therefore when the clock pulse arrives, FF1 is reset and Q1=0, not Q1=1.

FF2: before CP1, J=1 and K=0. Remember that J is tied to Q3 and K is tied to not Q3. Therefore when the clock pulse arrives, FF2 is set and Q2=1, and not Q2=0.

FF3: before CP1, J=1 and K=0. Therefore when the clock pulse arrives, FF3 is set; Q3=1 and not Q3=0.

The shift register is storing the following bits:

Q1	Q2	Q3
0	1	1

After CP2

From our preceding evaluation of the flip-flops after CP1, we can predict what will happen when the second clock pulse arrives:

FF1: after CP1, J=Q3=1 and K=not Q3=0. Therefore when the second clock pulse arrives, FF1 is set; Q1=1, and not Q1=0.

FF2: after CP1, J=Q1=0 and K=not Q1=1. Therefore when the second clock pulse arrives, FF2 is reset; Q2=0 and not Q2=1.

FF3: after CP1, J=Q2=1 and K=not Q2=0. Therefore when the second clock pulse arrives, FF3 is set; Q3=1 and not Q3=0. The binary number stored in the shift register is:

Q1	Q2	Q3
1	0	1

After CP3

From our preceding evaluation of the flip-flops after CP2, we can predict what will happen when the third clock pulse arrives.

FF1: after CP2, J=Q3=1 and K=not Q3=0. Therefore when the third clock pulse arrives, FF1 is set; Q1=1 and not Q1=0.

FF2: after CP2, J=Q1=1 and K=not Q1=0. Therefore when the third clock pulse arrives, FF2 is set; Q2=1 and not Q2=0.

FF3: after CP2, J=Q2=0 and K=not Q2=1. Therefore when the third clock pulse arrives, FF3 is reset; Q3=0 and not Q3=1. The shift regsiter is storing the following binary values:

Q1	Q2	Q3
1	1	0

Note that the contents of the parallel-load shift register are the same before CP1 and after CP3. Three clock pulses are needed in a 3-bit parallel-load shift register to recirculate the binary bits to their original positions.

After CP4
Same as "after CP1."

Q1	Q2	Q3
0	1	1

As we have seen, the data recirculates in the parallel-load shift register and is never lost. When we wish to load new data, the shift register should be cleared first.

Arithmetic Operations with Shift Registers

In this section we will be using shift registers to accomplish multiplication and division. By shifting a binary number one space to the right we divide the binary number by 2. Shifting the binary number to the left by one space is the same as multiplying the binary number by 2. Note that no recirculating data shift registers can be used when arithmetic operations are performed.

Binary Division with Shift Registers

To perform a binary division by 2, right shift registers are used. Now let's divide the binary 0110 by 2. (The decimal equivalent of 0110 is 6.)

1 After clearing the shift register, let's load the binary 0110. (The input to the shift right register can be either serial or parallel.) The shift register outputs become:

Q1	Q2	Q3	Q4
0	1	1	0

2 Now let's apply one clock pulse to the shift register. The shift register's contents become:

Q1	Q2	Q3	Q4
0	0	1	1

The binary number is now 0011. The decimal equivalent of 0011 is 3. Therefore by shifting the binary number one space to the right we have divided by 2.

Example

Use a right shift register to divide the decimal number 20 by 4.

Solution

Let's first convert the decimal number into its binary equivalent.

20 = 10100

Our binary number is composed of 5 bits. Therefore we'll need a 5-bit shift register to store the binary number.

Let's now load binary 10100 into a 5-bit shift register.

Q1	Q2	Q3	Q4	Q5
1	0	1	0	0

To divide by 4, two shifts to the right are required. Therefore we have to apply two clock pulses to the shift register. The shift register's contents become:

Q1	Q2	Q3	Q4	Q5
0	0	1	0	1

Let's convert binary 00101 to its decimal equivalent:

00101 = 5

By shifting the binary number two times to the right, we've divided by four.

Binary Multiplication with Shift Registers

Figure 3.15 shows a 4-bit left shift register. Note that the flip-flop's are inverted. This configuration assures a left shift operation. The operation of a left shift register is identical to that of a right shift register. The only difference is that one shifts right; the other shifts left.

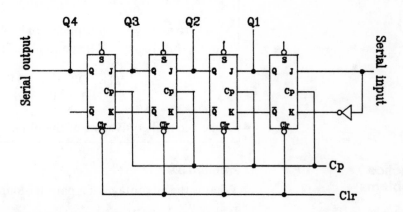

Figure 3.15. 4-bit left shift register (serial in-serial out)

Left shift registers are used to multiply a binary number by 2. Every time a binary number is shifted one position to the left, it is multiplied by 2.

Using serial input (we could have used a parallel-load left shift register as well), let's load the binary number 0111 (or decimal 7) into the left shift register of figure 3.15. (Assume that shift register was cleared before the loading process began.)

Four clock pulses are required to load binary 0111 into the shift register. Once the loading process is complete, the shift register's outputs become:

Q1	Q2	Q3	Q4
0	1	1	1

Now, let's apply one more clock pulse to multiply the binary number stored in the shift register by 2. The contents of the shift register become:

Q1	Q2	Q3	Q4
1	1	1	0

The binary number stored in the shift register is now 1110. 1110 equals 14 in the decimal number system. Therefore shifting the number to the left in effect multiplied it by 2.

Example
Use the left shift register of figure 3.15 to multiply the decimal number 3 by 4.

Solution
The binary equivalent of 3 is 11. Let's load binary 11 into the left shift register. Note that only two clock pulses are needed.

Q1	Q2	Q3	Q4
0	0	1	1

Now, to achieve a multiplication by 4, two clock pulses must be applied to the shift register. The contents of the shift register become:

Q1	Q2	Q3	Q4
1	1	0	0

The decimal equivalent of the binary number 1100 is 12.

Practice Problems

Problem 3.9
What is the advantage of a parallel-load shift register over a serial input shift register?

Solution

Parallel-load is asynchronous. Therefore a parallel-load shift register requires no clock pulses to load data. On the other hand a serial-load shift register requires as many clock pulses as there are bits to be loaded.

Problem 3.10

What causes the data to recirculate in the parallel-load shift register of figure 3.13? Explain.

Solution

The outputs Q and not Q of FF3 are fed back to the J and K inputs of FF1 respectively. This configuration saves the data. In other words, the feedback lines from FF3 to FF1 permit the reloading of the bits that exit FF3 into FF1.

Problem 3.11

How many clock pulses are required to recirculate the data in a 8-bit parallel-load shift register.

Solution

In an 8-bit parallel-load shift register, 8 clock pulses are required to recirculate the data.

Problem 3.12

Design a 4-bit parallel-load shift register. The data should recirculate. Use D flip-flops.

Solution

Figure 3.16 represents a 4-bit parallel-load shift register.

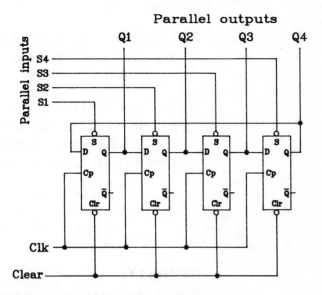

Figure 3.16. 4-bit parallel-load shift register

Problem 3.13

Load the decimal 12 into a 4-bit parallel-load shift register (with recirculating data). Describe this process step by step. (We will be loading decimal 12 into the shift register pictured in figure 3.16.

Solution

	Q1	Q2	Q3	Q4	
Step 2	0	0	0	0	(Clearing the shift register)
Step 3	1	1	0	0	(Loading binary 1100)

4-bit parallel-load shift register

1 To load the decimal 12 into the parallel-load shift register, we should start by converting the decimal 12 into its binary equivalent.

The binary equivalent of the decimal 12 is 1 1 0 0.

2 Before loading binary 1100 into the shift register, we have to clear the shift register. To clear the shift register the clr input line should be set low for a short moment.

Recall that the clr input line is low true input. Therefore a low input signal (or ground) on the clr input line would enable the input and the shift register would be cleared. A high input signal (or 5 volt) would disable the clr input line. (In normal operation of the shift register, the clr input line should be disabled.)

3 Once the shift register is cleared, binary 1100 can be loaded. To load binary 1100, we will be using the parallel inputs S1, S2, S3, and S4. The parallel inputs are low true. Therefore a low signal would enable the input.

To load binary 1100, FF1 and FF2 should be set (Q1=Q2=1), and FF3 and FF4 should be reset (Q3=Q4=0). Now let's set FF1 and FF2. To set FF1 and FF2, the asynchronous inputs S1 and S2 should be brought low for a short moment. S3 and S4 must remain high at all times.

After the loading process, the shift register's outputs are:

Q1	Q2	Q3	Q4
1	1	0	0

Problem 3.14

What decimal numbers are stored in the shift register after four clock pulses are applied to the shift register described in problem 3.13?

Solution

1 When the first clock pulse arrives, the binary bits stored in the shift register will be shifted one space to the right. Note that because of the recirculating data, the binary bit stored in FF4 is transfered to FF1.

After CP1

Q1	Q2	Q3	Q4
0	1	1	0

The decimal equivalent of 0110 is 6.

2 After CP2

When the second clock pulse arrives, the binary number stored in the shift register is again shifted to the right.

Q1	Q2	Q3	Q4
0	0	1	1

The decimal equivalent of binary 0011 is 3.

3 After CP3

Again, when the third clock pulse arrives, the contents of the shift register are shifted one space to the right.

Q1	Q2	Q3	Q4
1	0	0	1

The decimal equivalent of binary 1001 is 9.

4 After CP4

When the fourth clock pulse arrives, the contents of the shift register are shifted to the right.

Q1	Q2	Q3	Q4
1	1	0	0

The decimal equivalent of the binary 1100 is 12.

Four clock pulses are required to recirculate the data through a 4-bit parallel-load shift register. The contents of the shift register after CP4 are the same as those before the first clock pulse was applied.

Problem 3.15

Describe all steps necessary (design and operation) for dividing the decimal value 9 by 4. Explain.

Solution

To divide the decimal value 9 by 4, we'll follow these steps:

1 Design an appropriate shift register.

2 Load the binary equivalent into the shift register, and divide it by 4.

Step 1 contains the following substeps:

1 Convert decimal 9 into its binary equivalent.

2 Design a shift register that can load the binary equivalent of 9.

3 Is the shift register of step 2 capable of storing the result of the division by 4? If not make the necessary changes in the shift register to accommodate the result of the division.

Step 2 includes the following substeps:

1 Clear the shift register, and show its contents.

2 Load the binary equivalent of 9 into the shift register.

3 Shift the contents of the shift register two spaces to the right to divide by 4.

4 Obtain the binary result of the division.

5 Convert the binary result into its decimal equivalent.

Now let's implement the mentioned substeps to solve our problem.

Step 1
1 The binary equivalent of decimal 9 is 1001.

2 The binary number to be loaded (1001) consists of four bits. Therefore a 4-bit shift right register is needed. Recall that a shift register to be used in an arithmetic operation must not recirculate its data. A parallel-load or a serial load shift register could be used. In this problem a serial input shift register is used. The following shift register can be used to store binary 1001.

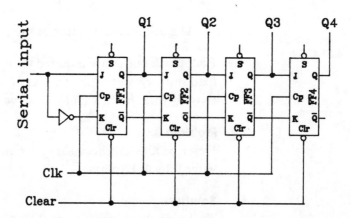

3 The result of the division (9 by 4) contains a fraction. Therefore the shift register of step 2 must be expanded to accommodate the fraction. We already know that two shifts to the right are required. Therefore we need two more flip-flops to the right of the shift register of step 2 to store the fractional part of the binary division. Otherwise when the binary number 1001 is shifted twice to the right, the rightmost two bits (01) will be lost. Remember that the two additional flip-flops that are going to be added represents the fractional part of the binary division. The final shift register follows:

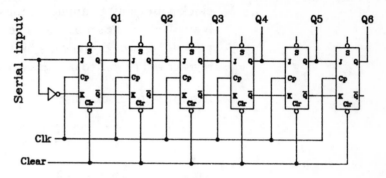

Q1 Q2 Q3 Q4: **Binary whole–part output**

Q5 Q6: **Binary fraction–part output**

Step 2

1 To clear the shift register, let's bring the clr input line low for a short period of time. Recall that the clr input is low true; in other words a low signal enables the clr input. In normal operation, the clr input line should be high (clr input is disabled by a high signal). The shift register's contents are:

Q1	Q2	Q3	Q4	.	Q5	Q6
0	0	0	0	.	0	0

2 Let's load binary 1001 into the shift register. To load the rightmost binary bit (LSB), we first bring the serial input line high and then apply a clock pulse. After the first clock pulse, the binary bit 1 has been loaded. The shift register's contents are now:

Q1	Q2	Q3	Q4	.	Q5	Q6
1	0	0	0	.	0	0

Now let's bring the serial input low and apply two clock pulses to load the two 0's that follow 1. The shift register's contents become:

Q1	Q2	Q3	Q4	.	Q5	Q6
0	0	1	0	.	0	0

Again let's bring the serial input line high and apply a clock pulse to load the last binary bit 1 (or MSB). The shift register's contents are:

Q1	Q2	Q3	Q4	.	Q5	Q6
1	0	0	1	.	0	0

The loading process is complete. The serial input line should be brought low for normal operation.

3 To shift the contents of the shift register by two spaces to the right, two clock pulses must be applied to the shift register. Recall that when no data is available in the serial input line, a binary 0 is loaded each time a clock pulse is applied. Each time a binary 0 is loaded, the contents of the shift register will be shifted one space to right. The shift register's contents are:

After the first clock pulse:

Q1	Q2	Q3	Q4	.	Q5	Q6
0	1	0	0	.	1	0

After the second clock pulse:

Q1	Q2	Q3	Q4	.	Q5	Q6
0	0	1	0	.	0	1

Note that the binary bits that cross the binary point (from Q4 to Q5), become part of the binary fraction.

After the first shift, we obtained the binary number 0100.10. After the second shift the binary number 0010.01 was obtained.

4 The binary result of the division is:

0010.01_2

5 Let's convert binary 0010.01 to its decimal equivalent:

$$0010.01_2 = (0 \times 2^3) + (0 \times 2^2) + (1 \times 2^1) + (0 \times 2^0) + (0 \times 2^{-1}) + (1 \times 2^{-2})$$
$$= 0 + 0 + 2 + 0 + 0 + 0.25$$
$$= 2.25_{10}$$

Experiment 5

Purpose   To build and operate a 4-bit right shift register.

Parts List

Quantity	Parts Description
1	Temporary board
1	5 v power supply
5	Resistors (200 ohms each)
4	Leds
1	Push button
2	Switches
2	7476 IC (Dual J-K flip-flops)
1	7404 IC (6 NOT gates)—(Hex Inverter)

Design Procedure

We will be experimenting with a 4-bit right shift register made of J-K flip-flops. The shift register has a serial input and a parallel output. Figure 3.17 represents a 4-bit right shift register. Note that the J-K flip-flops are negative-edge triggered; in other words only the high to low transition of the clock triggers the flip-flops. Q1, Q2, Q3, and Q4 represent the parallel outputs of the shift register.

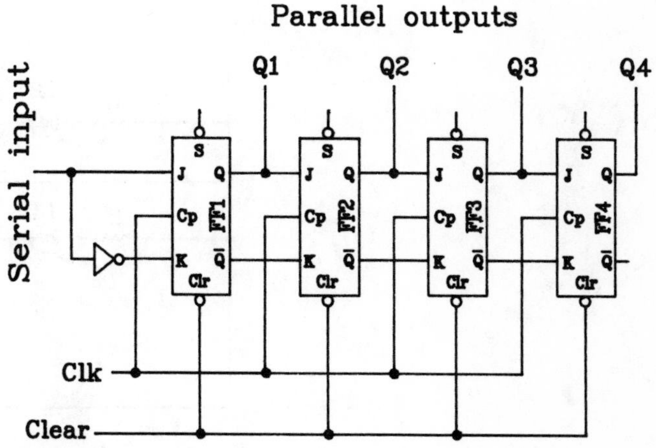

Figure 3.17. 4-bit right shift register

Operating Procedure

Wire the circuit shown in figure 3.18 (refer to the earlier experiments for wiring procedures and the internal description of the 7404 and 7476 ICs). Connect the battery (or the power supply) last.

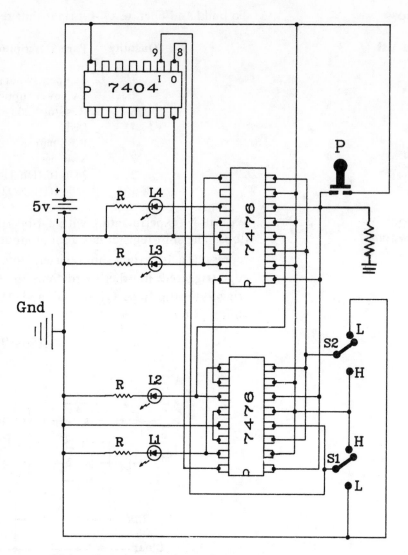

Figure 3.18. Wiring diagram--experiment 5

Circuit Description

Leds

L1, L2, L3, and L4 represent respectively Q1, Q2, Q3, and Q4. If an led is on, it represents binary 1. If the led is off, it represents binary 0.

Switches

S1: Serial input. When S1 is in the H (high) position, the serial input line is high. When S1 is in the L (low) position, the serial input line is low.

S2: Clear input. When S2 is in the L (low) position the clear input line is low, and the shift register is cleared. In normal operation, S2 should be in the H (high) position to disable the clear input line.

Push button P

Push the P push button to clock the circuit. Every time P is pushed one clock pulse is sent to the shift register.

Experimenting with the Shift Register

As an example, let's load binary 1101 into the shift register.

1 First, let's clear the shift register by putting the S2 switch in the low position. This causes pins 3 and 8 in both 7476 IC's to go low. Pins 3 and 8 represent the clr inputs. After step 1 all the leds (L1, L2, L3, and L4) should be off. If not, disconnect your power supply, and check your circuit.

Once the shift register is cleared (all leds are off), return the S2 switch to the H (or high) position. S2 should remain in the H position throughout the normal operation of the shift register. (S2 is brought to the L position *only* if the shift register is to be cleared.)

2 Let's load the LSB (least significant bit) of binary 1101 (or the rightmost binary bit). First, let's make the serial input line high (or 1) by setting switch S1 to the H (or high) position. Next, push the P push button once to trigger the shift register. L1 should be "on" and L2, L3, and L4 should be "off." The binary bit 1 is loaded.

L1 (= Q1)	L2 (= Q2)	L3 (= Q3)	L4 (= Q4)
on (= 1)	off (= 0)	off (= 0)	off (= 0)

3 Let's load the binary bit that is to the immediate right of the LSB (binary 0). First, switch S1 to the L (or low) position. Second, push P once. L1 is now "off," L2 is "on," L3 and L4 are "off." The binary bit 0 was loaded.

L1 (= Q1)	L2 (= Q2)	L3 (= Q3)	L4 (= Q4)
off (= 0)	on (= 1)	off (= 0)	off (= 0)

4 Let's load the third binary bit (from the right). First, switch S1 to the H position. Second, push P once. L1 is now "on," L2 is "off," L3 is "on," and L4 is "off." The binary bit 1 was loaded into the shift register.

L1 (= Q1)	L2 (= Q2)	L3 (= Q3)	L4 (= Q4)
on (= 1)	off (= 0)	on (= 1)	off (= 0)

5 Let's load the last binary bit (MSB). First, switch S1 should be in the H position. Second, push P once. L1, L2, and L4 are "on;" L3 is "off." The MSB (bit 1) was loaded into the shift register. The loading process of the binary number 1101 is finished.

L1 (= Q1)	L2 (= Q2)	L3 (= Q3)	L4 (= Q4)
on (= 1)	on (= 1)	off (= 0)	on (= 1)

Additional Practice

We have loaded the binary 1101. Now clear the shift register and load binary 1100. Use the shift register of figure 3.18 to divide binary 1100 by 4. Once binary 1100 is loaded, switch S2 to the low position and push P two times. The contents of the shift register should become 0011 which is 1100 divided by 4.

Problems

Problem 3.1

State four applications of shift registers.

Problem 3.2

Discuss the differences between serial-load shift registers and parallel-load shift registers.

Problem 3.3

Compute the number of flip-flops required to store a byte.

Problem 3.4

How many clock pulses are required to load 16 bits into a 16-bit serial-load shift register?

Problem 3.5
Design a 8 bit serial-load shift register using D flip-flops.

Problem 3.6
Design a four bit serial-load shift register using J-K flip-flops.

Problem 3.7

How many clock pulses are required to load 16 bits into a 16-bit parallel-load shift register?

Problem 3.8

Design an 8-bit parallel-load shift register using D flip-flops.

Problem 3.9

Design a circuit (shift register) that can hold three decimal digits.

Problem 3.10

Design a shift register that can execute the following decimal division: 18 by 4.

4

Decoders/Encoders

Introduction

In chapters 2 and 3 we learned about sequential circuits such as counters and shift registers. In this chapter and the following one, we'll learn how combinational circuits operate.

Combinatorial circuits are *decision making circuits*. In other words, when input information is given (usually a binary code), the combinational circuit will decide what measures are to be taken for a specific situation (output a certain binary code). The output(s) of a combinational circuit is always a function of the input(s). In other words, the input signal(s) to the combinational circuit will dictate what output signal(s) is to be generated. Because of this relation between the input and output signals, combinational circuits are often called *functional logic circuits*.

Combinational (or functional) logic circuits are made of logic gates such as AND, NOR, NOT,..etc. The most common combinational logic circuits are *decoders, encoders, multiplexers, demultiplexers, programmable logic arrays,* and *read only memories*.

There is no need to design these combinational circuits. They are found ready for use, as completely wired MSI (Medium Scale Integration) integrated circuits.

Figure 4.1 contrasts applications of functional logic circuits with those of sequential logic circuits.

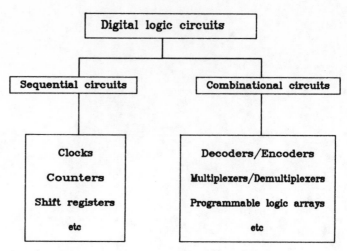

Figure 4.1. Contrast between sequential and combinational logic circuits.

Decoders

Definition

Basically, a *decoder* is a combinational logic circuit that will detect a particular binary number (or bit).

One of the simplest decoders is an OR gate. An OR gate can be used to detect the presence of the binary bit 1 given a binary number. Suppose that we were given two binary bits (inputs) and we wanted to detect the presence of binary 1. Suppose A and B were the inputs to be checked for the presence of binary bit 1 (inputs to the decoder), and Y is the output of the OR gate (or output of the decoder that will tell us if either one input or both inputs are high).

There are four possible input combinations (*see* figure 4.2).

(Inputs)		(Output)
A	B	Y
0	0	0
0	1	1
1	0	1
1	1	1

Figure 4.2. OR gate as a decoder

Note that whenever either A or B, or both A and B are 1, Y is 1. Therefore an OR gate detects the presence of binary 1 and can be considered a decoder.

Figure 4.3 represents a decoder that would detect the binary number 101.

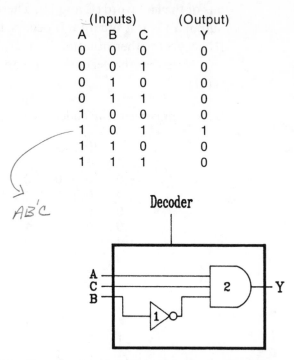

(Inputs)			(Output)
A	B	C	Y
0	0	0	0
0	0	1	0
0	1	0	0
0	1	1	0
1	0	0	0
1	0	1	1
1	1	0	0
1	1	1	0

$AB'C$

Figure 4.3. Decoder that detects the binary number 101

Notice that when the binary number 101 is sensed at the inputs (respectively ABC), Y is 1. The combinational circuit of figure 4.3 is indeed a decoder that detects the binary number 101.

A decoder may have multiple outputs as well as multiple inputs. Let's suppose now that we need a decoder that will tell if only the two leftmost binary bits of a half word (4 bit binary number=half word) are ones, and if only the two rightmost binary bits of the half word are ones. The decoder should be able to differentiate between three different situations.

Situation I
Only the two leftmost binary bits of the half word are ones.

Situation II
Only the two rightmost binary bits of the half word are ones.

Situation III
Neither situations I nor II are present.

Let's name our inputs A B C D, where A and B represent the two leftmost binary bits of the half word, and C and D represent the two rightmost binary bits of the half word.

There are three different situations to distinguish. Therefore at least two outputs are required to tell which situation is occuring. Let's name the outputs of the decoder Y1 and Y2. Output Y1 keeps track of the two leftmost binary bits of the half word (A and B), and Y2 keeps track of the two rightmost binary bits of the half word (C and D). Therefore if situation I occurs, the decoder's output Y1 is 1; if situation II occurs, the decoder's output Y2 is 1. If situation III occurs (neither situation I nor II), neither decoder outputs (Y1 or Y2) are one. Note that the decoder will ignore the case where all the inputs are high (1111).

Let's summarize the decoder operation in the following truth table:

(Inputs)				(Output)	
A	B	C	D	Y1	Y2
0	0	0	0	0	0
0	0	0	1	0	0
0	0	1	0	0	0
0	0	1	1	0	1
0	1	0	0	0	0
0	1	0	1	0	0
0	1	1	0	0	0
0	1	1	1	0	0
1	0	0	0	0	0
1	0	0	1	0	0
1	0	1	0	0	0
1	0	1	1	0	0
1	1	0	0	1	0
1	1	0	1	0	0
1	1	1	0	0	0
1	1	1	1	0	0

$Y' = ABC'D'$

The decoder that performs the above operation is represented in figure 4.4.

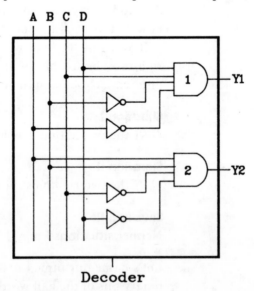

Decoder

Figure 4.4. Multiple output decoder

The design of a combinational circuit (truth tables, Karnaugh maps,..etc) was covered in Digital Circuits Vol. 2. If you encounter any difficulties designing combinational circuits in the following examples, review Digital Circuits Vol. 2.

Example

Design a decoder which outputs binary bit 1 when it detects the presence of more than two binary 1's in its four binary inputs.

The design procedure for this decoder consists of the following steps:

1 Truth table implementation

Let's name the inputs A B C D, and let's name the output Y. The truth table follows:

(Inputs)				(Output)
A	B	C	D	Y
0	0	0	0	0
0	0	0	1	0
0	0	1	0	0
0	0	1	1	0
0	1	0	0	0
0	1	0	1	0
0	1	1	0	0
0	1	1	1	1
1	0	0	0	0
1	0	0	1	0
1	0	1	0	0
1	0	1	1	1
1	1	0	0	0
1	1	0	1	1
1	1	1	0	1
1	1	1	1	1

The minterm Boolean expression is:

$$Y = \overline{A}BCD + A\overline{B}CD + AB\overline{C}D + ABC\overline{D} + ABCD$$

2 Karnaugh map

The Karnaugh map corresponding to the above truth table is:

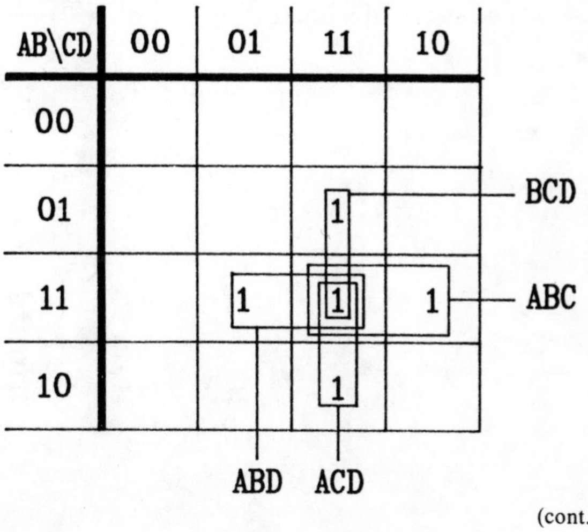

AB\CD	00	01	11	10	
00					
01			1		——— BCD
11		1	1	1	——— ABC
10			1		

ABD   ACD

(cont. on page 150)

3 Reduced minterm Boolean expression

Y = ABC + ABD + ACD + BCD

4 Decoder design

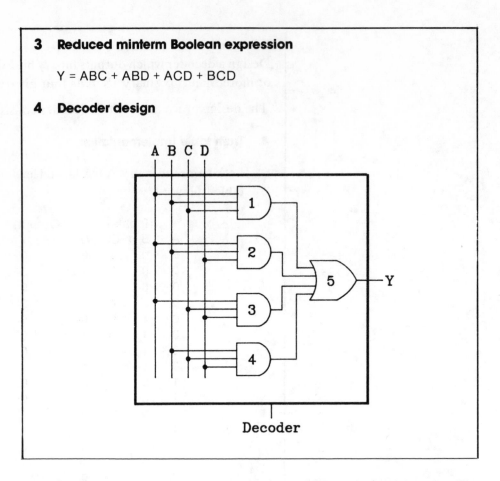

BCD to Decimal Decoder

One of the most commonly used decoders is the BCD to decimal decoder. The BCD to decimal decoder converts from 8421 BCD (Binary Coded Decimal) to the decimal number system. Recall that the 8421 BCD uses four bits to represent one decimal digit. Figure 4.5 represents the logic symbol for a BCD to decimal decoder.

Figure 4.6 represents a commercial BCD to decimal decoder. This commercial TTL (transistor to transistor logic) chip is known as the 7442 BCD to decimal decoder.

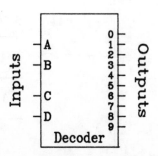

Figure 4.5. BCD to decimal decoder

The decoder consists of four parallel inputs (A, B, C, and D) and ten outputs (decimal digits 0 through 9). When an 8421 BCD number is entered (inputs), *one and only one* output is activated. *The output activated is the decimal equivalent of the 8421 BCD number entered.* Therefore the BCD to decimal decoder translates from the 8421 Binary Coded Decimal to the decimal number system.

Note that the A input represents the LSB (least significant bit) and has a weight of 1, B has a weight of 2, C has a weight of 4, and finally D represents the MSB (most significant bit) and has a weight of 8.

Let's suppose that the inputs and outputs of the decoder are high true. In other words a high signal (or 1) activates the inputs. Now, if for example we activate the A and C inputs, by applying 1 to both inputs A and C, and deactivate the B and D inputs by applying a 0 to both inputs B and D; the 8421 BCD number 0101 shows at the decoder inputs. This causes the 5 output to be activated; a high signal (or 1) will be present at output 5. (The decimal digit 5 is represented by 0101 in the 8421 BCD number system.) All other outputs (0, 1, 2, 3, 4, 6, 7, 8, and 9) are inactivated (a low signal is present at the outputs).

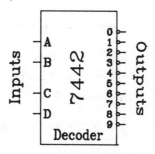

Figure 4.6. Commercial BCD to decimal decoder

The outputs of the 7442 IC are low true (note the bubbles at the outputs). In other words, an output of the 7442 IC is activated when a low signal (or logic 0) is floating at the output. On the contrary, the inputs to the 7442 IC are high true (note the absence of the bubbles at the inputs of the decoder). A high level signal (or logic 1) must be applied to activate an input.

Figure 4.7 summarizes the 7442 decoder's operation.

Inputs				Outputs										Output activated
D	C	B	A	0	1	2	3	4	5	6	7	8	9	
L	L	L	L	**L**	H	H	H	H	H	H	H	H	H	0
L	L	L	H	H	**L**	H	H	H	H	H	H	H	H	1
L	L	H	L	H	H	**L**	H	H	H	H	H	H	H	2
L	L	H	H	H	H	H	**L**	H	H	H	H	H	H	3
L	H	L	L	H	H	H	H	**L**	H	H	H	H	H	4
L	H	L	H	H	H	H	H	H	**L**	H	H	H	H	5
L	H	H	L	H	H	H	H	H	H	**L**	H	H	H	6
L	H	H	H	H	H	H	H	H	H	H	**L**	H	H	7
H	L	L	L	H	H	H	H	H	H	H	H	**L**	H	8
H	L	L	H	H	H	H	H	H	H	H	H	H	**L**	9
H	L	H	L	H	H	H	H	H	H	H	H	H	H	none
H	L	H	H	H	H	H	H	H	H	H	H	H	H	none
H	H	L	L	H	H	H	H	H	H	H	H	H	H	none
H	H	L	H	H	H	H	H	H	H	H	H	H	H	none
H	H	H	L	H	H	H	H	H	H	H	H	H	H	none
H	H	H	H	H	H	H	H	H	H	H	H	H	H	none

Figure 4.7. Truth table for the 7442 decoder

Recall that the 8421 BCD consists of binary 0000 through 1001. Therefore when a non-8421 BCD number is entered, the 7442 IC doesn't recognize it, and no output is activated (*see* "output activated = none" in figure 4.7).

The logic diagram of the 7442 IC decoder is shown in figure 4.8.

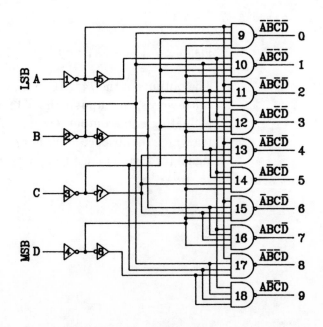

Figure 4.8. Logic diagram of the 7442 decoder

Let's test the logic circuit of figure 4.8 by tracing the randomly chosen 8421 BCD number 0110 (=DCBA) through the gates. Note that D represents the MSB (most significant bit), and A the LSB (least significant bit).

Gate 9's output is H

The input logic function to gate 9 is DCBA (*see* figure 4.8). Given that 0110=DCBA, gate 9's input is 1001. Therefore the output of gate 9 is 1 (or H).

Recall that for a NAND gate's output to be low (or 0), all of the inputs to the NAND gate must be high. In other words, if at least one input to the NAND gate is low (or 0), the output of the NAND gate is high (or 1).

The binary input to gate 9 (1001) was obtained as follows. The inputs DCBA to the decoder are respectively 0110. (0110 is the randomly choosen binary number used to test the decoder circuit of figure 4.8.) From figure 4.8, we see that the inputs to gate 9 are as follows:

$$\text{Gate 9's inputs: } \overline{D}\ \overline{C}\ \overline{B}\ \overline{A}$$

Therefore, the binary inputs (0 1 1 0=D C B A) are complemented when they arrive at gate 9 (1 0 0 1=$\overline{D}\ \overline{C}\ \overline{B}\ \overline{A}$). Gate 9's inputs are always the complement of the binary number that is entered into the 7442 decoder.

Gate 10's output is H

The input function to gate 10 is $\overline{D}\ \overline{C}\ \overline{B}\ A$. Therefore when a binary number DCBA is entered into the decoder, the three bits D, C, and B are complemented when entering gate 10.

Thus, when 0110 is entered into the decoder, it becomes 1000 when it enters gate 10. Therefore the output of gate 10 is high (or 1).

Gate 11's output is H

The input function to gate 11 is $\overline{D}\ \overline{C}\ B\ \overline{A}$. Therefore when a binary number DCBA is entered into the decoder, the bits D, C, and A are complemented when entering gate 11.

Thus, when 0110 is entered into the decoder, it becomes 1011 when it enters gate 11. Therefore the output of gate 11 is high (or 1).

Gate 12's output is H

The input function to gate 12 is $\overline{D}\ \overline{C}\ B\ A$. Therefore the bits D and C are inverted when entering gate 12. Thus, when 0110 is entered into the decoder, binary 1010 is present at gate 12. Therefore the output of gate 12 is high (or 1).

Gate 13's output is H

The input function to gate 13 is $\overline{D}\ C\ \overline{B}\ \overline{A}$. Bits D, B, and A are complemented before entering gate 13. Thus the binary number present at gate 13 when the binary 0110 is entered into the decoder is 1101. The output of gate 13 is high (or 1).

Gate 14's output is H

The input function to gate 14 is $\overline{D}$ C $\overline{B}$ A. The bits D and B are complemented when entering gate 14. Thus, the binary 1100 is present at gate 14 when 0110 is entered into the decoder. The output of gate 14 is then high (or 1).

Gate 15's output is L (Activated)

The input function to gate 15 is $\overline{D}$ B C $\overline{A}$. Bits D and A are complemented before entering gate 15. Thus when the binary number 0110 is entered into the decoder, gate 15 sees it as 1111. When all the inputs to a NAND gate are high (or 1), its output is low.

The output of gate 15 represents decimal 6. Because gate 15 was activated by entering the binary number 0110 into the decoder, we conclude that the decimal equivalent of the 8421 BCD number 0110 (entered into the decoder) is the decimal 6. Using the table in figure 4.7, we arrive at the same result.

Gate 16's output is H

The input function to gate 16 is $\overline{D}$ C B A. The binary bit D is complemented before entering gate 16. Therefore when 0110 is entered into the decoder, it is present at gate 16 as 1110. The output of gate 16 when 0110 in entered into the decoder is high (or 1).

Gate 17's output is H

The input function to gate 17 is D $\overline{C}$ $\overline{B}$ $\overline{A}$. Therefore the binary bits C, B, and A are complemented before entering gate 17. When 0110 is entered into the decoder, 0001 results at gate 17. When 0001 is entered into gate 17 (NAND gate), the output is high (or 1).

Gate 18's output is H

The input function to gate 18 is D $\overline{C}$ $\overline{B}$ A. Bits C and B are complemented before entering gate 18. Therefore when 0110 is entered into the decoder, 0000 results at gate 18. Thus the output of gate 18 when 0110 is entered into the decoder is high (or 1).

Example

Using the logic circuits of figure 4.8, which NAND gate will be brought low when binary 0100 is entered into the decoder.

Solution

A NAND gate's output is low only if all its inputs are high. Therefore when 0100 is entered, the NAND gate to be brought low must see 0100 as 1111. Therefore the D, B, and A bits must be complemented before entering the NAND gate. The logic function of this gate should be $\overline{D}$ C $\overline{B}$ $\overline{A}$. Gate 13's (*see* figure 4.8) logic input function is $\overline{D}$C $\overline{B}\overline{A}$. Therefore when 0100 is entered into the decoder, 1111 results at the input of gate 13. Thus gate 13's output will be low (or 0). (Use the table in figure 4.7 to verify the results.

**BCD to 7
Segment
Decoder**

One of the most popular decoders is the 7447. The 7447 BCD to 7 *segment decoder/driver* is used to convert from 8421 BCD code (inputs) to 7 segment code (outputs). The word "driver" was added to the 7447 decoder because the 7447 decoder's outputs drive a 7 segment decimal display (readout). A 7 segment display as shown in figure 4.9.

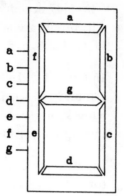

Figure 4.9. 7 segment display

The 7 segment display consists of 7 segments labeled "a" through "g." Each segment contains a display component. Many types of display components are used. Some of the most commonly used are the gaz discharge tube, liquid crystal display (or LCD), and light emitting diode (or LED).

When a segment is activated, it glows. If for example segments a, b, c, d, and g are activated, the decimal 3 would be displayed. The decimal numbers 0 through 9 and some alphabetical letters could be displayed with the 7 segment display. A commercial 7447 BCD to 7 segment decoder/driver is shown in figure 4.10.

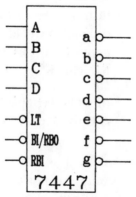

Figure 4.10. 7447 BCD to 7 segment decoder/driver

The 7447 decoder/driver accepts (inputs) four binary bits in 8421 standard BCD code and generates a 7 bit code that drives the 7 segment decimal display of figure 4.9. The 7 bit code generated by the 7447 decoder/driver (if the decoder/driver is connected to a 7 segment display) lights the decimal equivalent of the 8421 BCD code entered into the 7447 decoder/driver.

If for example the 8421 BCD number 0110 is entered into the 7447 decoder/driver, the decoder would generate a 7 bit code that activates segments f, e, d, c, and g. Therefore decimal 6 is displayed. Note that decimal 6 is the decimal equivalent of the 8421 BCD number 0110.

The 7447 decoder/driver consists of 7 inputs and 7 outputs (*see* figure 4.10). The inputs are labeled A, B, C, D, LT, BI/RBO, and RBI. The outputs are labeled a, b, c, d, e, f, and g.

Inputs LT, BI/RBO, RBI, and all of the outputs are low true (note the bubbles). In other words the inputs LT, BI/RBO, and RBI are activated when a low signal (or logic 0) is applied; outputs a, b, c, d, e, f, and g are active when low signals (or logic 0) are present at those outputs.

Inputs A, B, C, and D are high true. In other words, when a high logic signal (or logic 1) is applied, the inputs (A, B, C, and D) are activated.

Inputs A, B, C, and D are used to enter the 8421 BCD number to be decoded into a 7 segment code, where input A is the LSB (least significant bit), and D is the MSB (or most significant bit). If for example the 8421 BCD number 0110 is to be entered into the decoder, a low signal (or logic 0) should be applied to inputs A and D, and a high signal (or logic 1) should be applied to inputs B and C.

Input LT (lamp test) is used to check if all the segments are good. When a low signal (or logic 0) is applied to the LT input, all the lamps (of the 7 segment display) are on.

Input BI/RBO (blanking input/ripple blanking output) is used to blank (or turn off) the undesired 0's. If for example we have a four digit display, the decimal 35 would be displayed as 0035 when BI/RBO is disabled (high signal is applied to the BI/RBO input), The two 0's to the left of 35 are suppressed when BI/RBO is activated (a low signal is applied to the BI/RBO input).

Input RBI (ripple blanking input) is used to control the intensity of the display. A variable duty cycle is applied to the RBI input to vary the display intensity.

Outputs a, b, c, d, e, f, and g are used to drive the 7 segment display. When an output is not activated, the output floats high (or logic 1). If the output is activated, the output floats low (or logic 0). If for example the BCD number 0101 is entered into the 7447 decoder/driver (with the LT, BI/RBO, and RBI disabled), outputs a, f, g, c, and d all float low (or logic 0), and outputs b and e will float high. If the 7 segment decoder/driver is connected to a 7 segment display, the decimal number 5 will be displayed.

Figure 4.11 represents a truth table for the 7447 decoder/driver.

Inputs				Outputs							Decimal
D	C	B	A	a	b	c	d	e	f	g	displayed
0	0	0	0	0	0	0	0	0	0	1	0
0	0	0	1	1	0	0	1	1	1	1	1
0	0	1	0	0	0	1	0	0	1	0	2
0	0	1	1	0	0	0	0	1	1	0	3
0	1	0	0	1	0	0	1	1	0	0	4
0	1	0	1	0	1	0	0	1	0	0	5
0	1	1	0	1	1	0	0	0	0	0	6
0	1	1	1	0	0	0	1	1	1	1	7
1	0	0	0	0	0	0	0	0	0	0	8
1	0	0	1	0	0	0	1	1	0	0	9
1	0	1	0	1	1	1	0	0	1	0	N/A
1	0	1	1	1	1	0	0	1	1	0	N/A
1	1	0	0	1	0	1	1	1	0	0	N/A
1	1	0	1	0	1	1	0	1	0	0	N/A
1	1	1	0	1	1	1	0	0	0	0	N/A
1	1	1	1	1	1	1	1	1	1	1	N/A

Figure 4.11. Truth table for the 7447 decoder/driver

Note that when the logic output is 0 (low), its corresponding display segment is *on*, and when the logic output is 1 (or high), its corresponding display segment is *off*. If for example the outputs of a 7447 decoder/driver are 0 0 0 1 1 1 1 (respectively a b c d e f g), then display segments a, b, and c are on; and segments d, e, f, and g are off. Therefore the decimal 7 is displayed.

When a non BCD number is entered into the 7447 decoder/driver, some outputs are activated but the segments displayed are irrelevant to the decimal number system.

Example

What outputs of the 7447 decoder/driver are activated when the 8421 BCD number 0010 is entered?

Solution

When 0010 is entered, the output code generated is 0010010. Therefore outputs a, b, d, e, and g are activated. If the decoder is wired to a 7 segment display, the decimal number 2 would be displayed.

Practice Problems

Problem 4.1

Design a decoder that detects the binary number 010.

Solution

Let's generate a truth table for the decoder. First we will name the inputs to the decoder A, B, and C, and the output from the decoder Y. The truth table follows:

Inputs			Outputs
A	B	C	Y
0	0	0	0
0	0	1	0
0	1	0	1
0	1	1	0
1	0	0	0
1	0	1	0
1	1	0	0
1	1	1	0

Decoder

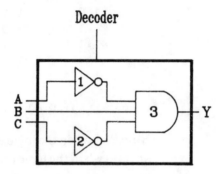

When the binary 010 is sensed at the inputs (respectively ABC), the output of gate 3 is 1. Therefore the combinational circuit depicted above detects binary 010. (Refer to volume 2 for more details about combinational design.)

Problem 4.2

Design a decoder that detects the presence of exactly two binary 1's in four inputs. In other words, when exactly two inputs are high, only then should the output of the decoder be 1.

Solution

Let's name the inputs to the decoder A, B, C, and D; and name the output Y.

1 Create the truth table

Inputs				Output
A	B	C	D	Y
0	0	0	0	0
0	0	0	1	0
0	0	1	0	0
0	0	1	1	1
0	1	0	0	0
0	1	0	1	1
0	1	1	0	1
0	1	1	1	0
1	0	0	0	0
1	0	0	1	1
1	0	1	0	1
1	0	1	1	0
1	1	0	0	1
1	1	0	1	0
1	1	1	0	0
1	1	1	1	0

The minterm Boolean expression is:

$$Y = \overline{A}\,\overline{B}CD + \overline{A}B\overline{C}D + \overline{A}BC\overline{D} + A\overline{B}\,\overline{C}D + A\overline{B}C\overline{D} + AB\overline{C}\,\overline{D}$$

2 Create a Karnaugh map

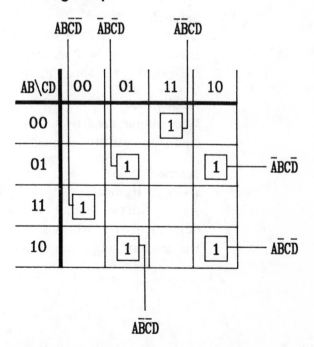

No two cells are adjacent in the Karnaugh map. Therefore no reduction is possible. All the terms found in the minterm Boolean expression of step 1 are used to implement the combinational circuit (decoder).

3 Design the decoder

From step 2 we arrive at (*see* Digital Circuits: Vol. 2).

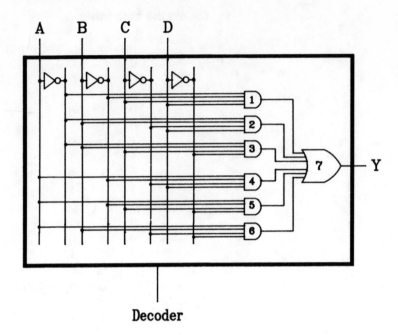

Decoder

Problem 4.3

Which output is activated when the binary number 1001 is entered into a 7442 decoder?

Solution

Using the truth table of figure 4.7, we see that when the binary number 1001 is entered into the 7447 decoder; output 9 is activated. Note that 9 is the decimal equivalent of the binary number 1001.

Problem 4.4

What outputs are activated by the BCD to 7 segment decoder/driver when the binary number 0101 is entered into the decoder.

Solution

Using the truth table of figure 4.11, we see that when the binary number 0101 is entered into the decoder/driver, outputs a, c, d, f, and g are activated. (The output is 0100100 which is respectively a, b, c, d, e, f, and g; recall that in low true logic an output is said to be activated if it is low.)

Problem 4.5

What decimal number is displayed when the binary number 0011 is entered into the following circuit:

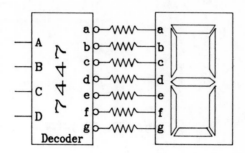

Solution

When the binary 0011 (respectively DCBA) is entered, led segments a, b, c, d, and g are activated. Therefore a 3 appears.

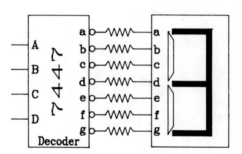

Encoders

Like decoders, *encoders* are also combinational logic circuits. Encoders are the opposite of decoders. While a decoder detects a specific code (or binary number), an encoder generates a specific code (or binary number).

Encoders are used in many applications. They are especially useful for translating keyboard signals (input) to the BCD number system (output). When a particular key is pressed, the encoder generates a specific code that is recognized by the CPU.

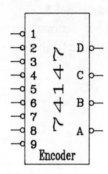

Figure 4.12. The 74147 encoder

One popular encoder is the 74147 encoder (*see* figure 4.12). The 74147, also called a *10-line-to-4-line priority encoder*, translates from the decimal number system to the 8421 BCD number system. The 74147 has 9 inputs (1 through 9) and 4 outputs (D, C, B, A). Note that the inputs and outputs of the 74147 encoder are low true. (Note the bubbles at the inputs and outputs). In other words, when a low signal is applied to an input, the input is activated. Similarly, when a low signal (or logic 0) is present at one output, this output is activated.

Figure 4.13 represents the logic diagram of the 74147 encoder.

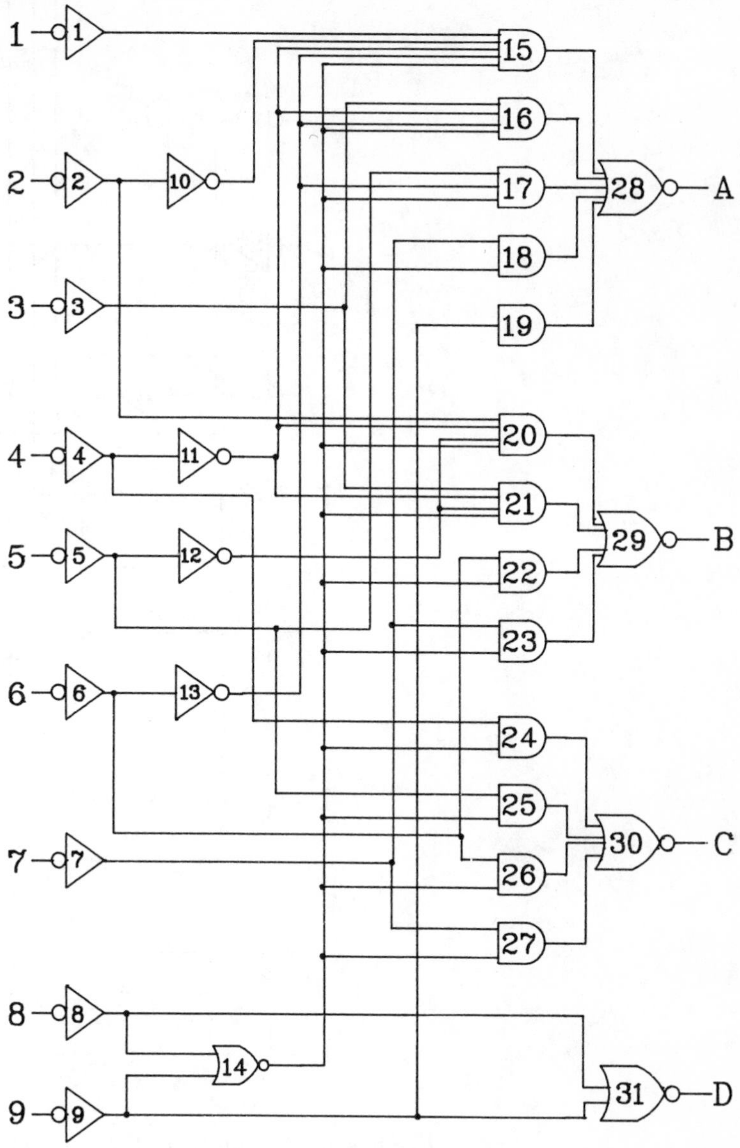

Figure 4.13. Logic diagram of the 74147 encoder

From figure 4.13 we note that the outputs of the 74147 encoder are inverted as they are leaving the encoder (bubbles at the outputs of gates 28 through 31). The input signals are also inverted when they enter the 74147 encoder (bubbles at the inputs of gates 1 through 9). These bubbles at the inputs and outputs of the 74147 encoder cause the logic of the combinational circuit to be *low true*.

If for example a low logic input signal (or logic 0) is applied to input 1, and high logic input signals are applied to inputs 2 through 9 (*see* figure 4.14), the A output will be activated and a low logic signal (or logic 0) will be present at output A. Let's follow the above signals through the gates of the logic diagram in figure 4.14.

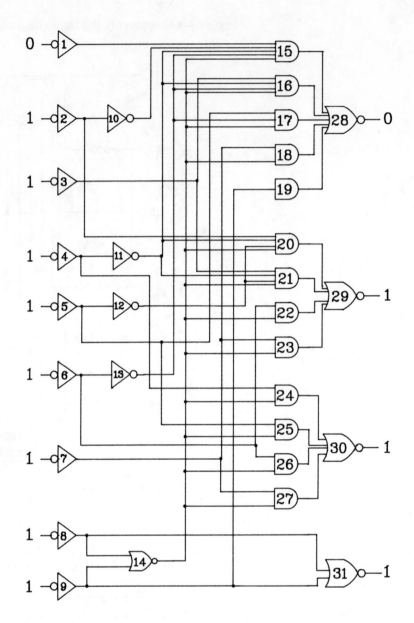

Figure 4.14. Operation of the 74147 encoder

When applying a low signal to input 1, and high signals to inputs 2 through 9, the following gate logic levels result:

	Gate #	Input(s) logic	Output logic
(NOT)	1	0	1
	2	1	0
	3	1	0
	4	1	0
	5	1	0
	6	1	0
	7	1	0
	8	1	0
	9	1	0
	10	0	1
	11	0	1
	12	0	1
	13	0	1
	14	0	1
(AND)	15	11111	1
	16	0111	0
	17	011	0
	18	01	0
	19	0	0
	20	0111	0
	21	0111	0
	22	01	0
	23	01	0
	24	01	0
	25	01	0
	26	01	0
	27	01	0
(NOR)	28	10000	0 (A)
	29	0000	1 (B)
	30	0000	1 (C)
	31	00	1 (D)

Figure 4.15. Gate logic levels when input 1 is activated in figure 4.14

The output of the 74147 encoder when a low signal is applied to input 1 and a high signal is applied to inputs 2 through 9 is 1110 (DCBA respectively).

The truth table for the 74147 operation is given in figure 4.16.

```
        (Inputs)              (Outputs)
    1 2 3 4 5 6 7 8 9        D C B A
    L H H H H H H H H        H H H L
    X L H H H H H H H        H H L H
    X X L H H H H H H        H H L L
    X X X L H H H H H        H L H H
    X X X X L H H H H        H L H L
    X X X X X L H H H        H L L H
    X X X X X X L H H        H L L L
    X X X X X X X L H        L H H H
    X X X X X X X X L        L H H L
    H H H H H H H H H        H H H H
```

L=Low logic level   H=High logic level   X=Don't care

Figure 4.16. Truth table for the 74147 encoder

From figure 4.16 we see that when input 1 is activated (a low signal was applied to the 1 input), the code generated by the 74147 encoder is 1110 (respectively DCBA).

When input 2 is activated (second line in the table of figure 4.16), the binary code generated by the 74147 encoder is 1101 (respectively DCBA). Output B was activated.

"X" marked the 1 input line. An X is a don't care state which means that either a low or high signal could have been used. Because the decoder has a *priority characteristic*, the binary code generated corresponds to the highest input number activated. In other words, if the decimal input numbers 1 and 2 are activated simultaneously, the 74147 encoder will generate a binary code that corresponds to the highest decimal; in our case the encoder would generate a code corresponding to the decimal 2.

When input 3 is activated (third line in the table of figure 4.16), the binary code generated by the 74147 encoder is 1100 (respectively DCBA). Outputs A and B were activated (low logic). Again an "X" marks inputs 1 and 2. The "X" in inputs 1 and 2 means that if these inputs (1 or 2 or both) were activated while input 3 was activated, the 74147 encoder would generate a code for the decimal number 3 and ignore inputs 1 and 2. To prove the priority characteristic of the 74147 encoder, activate (by applying low signals) inputs 1, 2, and 3 and follow the signals throughout the gates in figure 4.14. The code generated would be that of the highest decimal number activated. (In our case the code generated would be 1100 which corresponds to the decimal number 3).

Complete the study of the 74147 encoder by examining the rest of the lines in the truth table of figure 4.16 in the same way we examined the three first lines. You can use the logic diagram of figure 4.13 to prove the logic operation of the 74147 encoder by following signals throughout the gates as we did in figure 4.14.

We said in the beginning of this section that the 74147 encoder translates from the decimal number system to the 8421 BCD number system. From the truth table of figure 4.15 we note that the code generated by the 74147 when a particular decimal input number is activated doesn't correspond to its equivalent 8421 BCD number! For example, when input 1 was activated, the code generated by the 74147 encoder is 1110 (which corresponds to HHHL in figure 4.16. 1110 is not the 8421 BCD number equivalent of the decimal number 1!

Now, if we complement the code generated by the 74147 encoder when the decimal input 1 was activated we arrive at 0001 (0001 is the complement of 1110.) 0001 is the 8421 BCD equivalent of the decimal number 1.

If we complement the code generated by the 74147 encoder (when the decimal input 2 is activated, *see* figure 4.16), we arrive at 0010. 0010 is the 8421 BCD number that corresponds to the decimal 2.

If we continue to complement the codes generated by the 74147 encoder when other decimal inputs are activated, their 8421 BCD equivalents are indeed obtained.

We conclude that the 74147 encoder generates the complement of the 8421 **BCD number system.(This complement is caused by the bubbles at the outputs of gates 28, 29, 30, and 31 in figure 4.13).** In order for the 74147 encoder to generate the 8421 BCD number equivalent of the decimal input activated, the output code must be complemented by adding negative NOT gates (negative NOT gates are NOT gates with a bubble at the input instead of a bubble at the output) to the outputs of the 74147 encoder as follows:

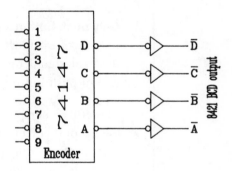

Figure 4.17. 74147 encoder with negative NOT gates

A practical application of figure 4.17 is shown in problem 4.9.

Example

What binary code number is generated when push buttons a, b, and c, are pressed one by one in the following push button encoder.

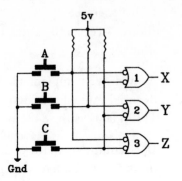

Figure 4.18. Push button encoder

Solution

1 No push button is pressed.

The bubbles at the inputs of gates 1, 2, and 3 (*see* figure 4.18) *invert* the input signals before they enter gates 1, 2, and 3. When no push button is pressed, 5 volts (or logic 1) is present at all the inputs of gates 1, 2, and 3 (*see* figure 4.18). The 5 volt signal (or logic 1) is complemented by the bubbles at the inputs of gates 1, 2, and 3. Therefore the 5 volt signal (or logic 1) becomes ground (or logic 0) once it enters gates 1, 2, and 3. The outputs X, Y, and Z of gates 1, 2, and 3 respectively are logic 0 when no push button is pressed.

X Y Z = 0 0 0 (When no push button is pressed)

2 Only push button A is pressed.

From figure 4.18 we see that one side of push button A is connected to the 5 volt signal, one input to gate 1 and one input to gate 3. The other side of push button A is connected to ground.

Therefore when push button A is pressed, a ground signal (also called logic 0 or 0 volt) propagates to one input of gate 1 and one input of gate 3. *Note that a 5 volt signal will always "disappear" when connected to ground.* Now, the ground signal (or logic 0) is inverted by the bubbles in the inputs of gates 1 and 3, thereby becoming logic 1. Therefore the output of gates 1 and 3 (respectively X and Z) are logic 1.

X Y Z=1 0 1 (When only push button A is pressed)

(cont. on page 169)

3 Only push button B is pressed.

From figure 4.18 we see that push button B is connected to the 5 volt signal and to one input of gate 2. The other side of B is connected to ground (or logic 0). When push button B is pressed, a ground signal (or logic 0) propagates through one input of gate 2. This ground signal (issued from push button B) is inverted by the bubble when it enters gate 2 and becomes logic 1. Gate 2's output (Y) is consequently logic 1.

X Y Z=0 1 0 (When only push button B is pressed)

4 Only push button C is pressed.

From figure 4.18 we see that one side of push button C is connected to the 5 volt signal and to one input of gates 1, 2, and 3. The other side of push button C is connected to ground (or logic 0). Therefore when push button C is pressed, a low signal (or ground) is sent to one input of all the gates. The low signals (issued from push button C) are inverted by the bubbles when entering gates 1, 2, and 3 and become high signals (or logic 1). The outputs of gates 1, 2, and 3 are consequently high (or logic 1).

X Y Z=1 1 1 (When only push button C is pressed)

Summary

Push button pressed	Code generated		
	X	Y	Z
none	0	0	0
A	1	0	1
B	0	1	0
C	1	1	1

Practice Problems

Problem 4.6

State the differences between encoders and decoders.

Solution

A decoder is a combinational logic circuit that will detect the presence of a specific binary number. The inputs to the decoder check for a specific binary number; the output from the decoder indicates the presence or absence of that specific number. Examples of decoders include the 7442 BCD to decimal decoder, and 7447 BCD to 7 segment decoder/driver.

An encoder is a combinational logic circuit that will generate a a specific binary code (output) when a specific input is activated. In other words, each input to the encoder corresponds to a binary code, and when one particular input is activated, its corresponding binary code is generated (by the encoder). The 74147 decimal to BCD priority encoder is an example of an encoder.

The main difference between decoders and encoders is that decoders detect specific binary codes, while encoders generate specific codes.

Problem 4.7

If inputs 2 and 5 of a 74147 encoder are activated with a low logic signal (*see* figure 4.19), what output code is generated and why?

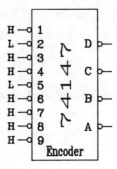

Figure 4.19. 74147 encoder with inputs 2 and 5 activated

Solution

Because the 74147 encoder has a priority feature, the 74147 encoder generates a code for the highest input number activated. In our case, both inputs 2 and 5 were activated. Therefore the 74147 encoder will generate a code that corresponds to decimal 5. From figure 4.16 we see that the code generated (output) by the encoder when decimal 5 was activated is 1010 (respectively DCBA).

Problem 4.8

What binary codes are generated when push buttons A, B, and C (*see* figure 4.20) are pressed one by one in the following push button encoder?

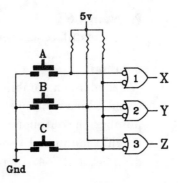

Figure 4.20. Push button encoder

Solution

Using the same techniques as in the example of figure 4.18 we arrive at:

When push button A is pushed, the binary code 100 is generated. When push button B is pushed, the binary code 011 is generated. When push button C is pushed, the binary code 111 is generated.

Problem 4.9

Change the wiring of the combinational circuit of problem 4.8 to obtain the following encoder operation:

	Push button pressed	Code generated
Step 1	none	0 0 0 (X Y Z)
Step 2	A	1 1 0 (X Y Z)
Step 3	B	0 1 1 (X Y Z)

Solution

1 All the input signals to gates 1, 2, and 3 should be high when no push button (either A or B) is pushed. This high signal (5v) becomes low when it enters the gates. (The bubbles at the gates's input invert the incoming signals). Therefore the output from each of the gates is low (or 0v) when no push button is pushed.

2 Push button A should be tied to gates 1 and 2, so that when it is pressed, a low signal will appear at the entrance of gates 1 and 2. The bubbles at the inputs invert signals before they enter the gates. Therefore when push button A is pushed, a low signal is generated. This low signal is inverted by the bubbles at the inputs of gates 1 and 2, causing the signal to go high. Therefore the outputs from gates 1 and 2 are high when push button A is pushed.

3 Push button B should be tied to gates 2 and 3, so that when push button B is pressed, low signals propagate to gates 2 and 3. Therefore the outputs from gates 2 and 3 are high.

The combinational circuit must be designed as follows (*see* figure 4.21):

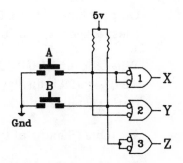

Figure 4.21. Push button encoder

Problem 4.10

What decimal number appears on the led when a low signal is applied simultaneously to inputs 5 and 9 (*see* figure 4.22)? Explain.

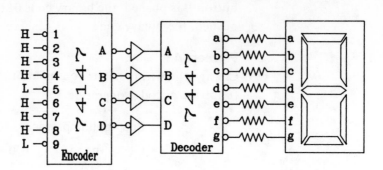

Figure 4.22. Led display

Solution

When a low signal is applied to the inputs of the 74147 encoder, the inputs are activated. In this case both inputs 5 and 9 are activated. Because the 74147 encoder has a priority feature, the encoder generates a code for the highest number activated. Therefore (from the table of figure 4.16) the 74147 encoder generates the following code:

Output of the 74147 encoder is 0 1 1 0 (Respectively D C B A)

The NOT gates invert the code generated by the 74147 encoder before it enters the 7447 BCD to 7-segment decoder. Therefore the binary code that enters the 7447 BCD to 7-segment decoder is:

Input code to the 7447 decoder is 1 0 0 1 (Respectively D C B A)

When the binary code 1001 (respectively DCBA) enters the 7447 BCD to 7-segment decoder, the decoder activates segments a, b, c, f, and g. (Refer to the table in figure 4.11). Recall the segments are low true and that a low signal is required to activate them. The output of the 7447 BCD to 7-segment decoder follows:

Output of the 7447 BCD to 7-segment decoder is 0 0 0 1 1 0 0
(Respectively a, b, c , d, e, f, and g)

When the binary 0 0 0 1 1 0 0 (respectively abcdefg) enters the 7-segment led, the decimal 9 is displayed (*see* figure 4.23). (Refer to figures 4.9 and 4.11 for further details about the 7447 decoder/driver and the 7-segment led operation.)

Summary of Operation

1 Inputs 9 and 5 of the 74147 are activated.

2 The 74147 priority encoder generates a code for input 9.

3 The code generated by the 74147 is complemented by the NOT gates between the 74147 encoder and the 7447 decoder/driver.

4 The complement of the code generated by the 74147 priority encoder enters the 7447 decoder/driver.

5 The 7447 decoder/driver activates segments a, b, c, f, and g.

6 When segments a, b, c, f, and g are activated, the decimal number 9 is displayed (*see* figure 4.23).

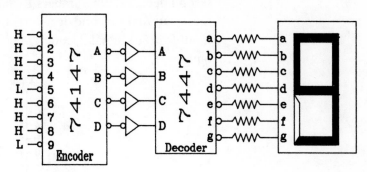

Figure 4.23. Summary of operation

Experiment 6

Purpose

To test and operate the 74ALS139 decoder.

Parts List

Quantity	Parts Description
1	Temporary board
1	5v power supply
4	Resistors (200 ohms each)
4	Leds
3	Switches
1	7404 IC (6 inverters)
1	74ALS139 IC (2-line to 4-line decoder)

74ALS139 Description

The 74ALS139 is a dual 2-line to 4-line decoder. By dual we mean that the 74ALS139 contains two decoders in a single chip. By 2-line to 4-line we mean that each decoder consists of 2 inputs and 4 outputs respectively.

The 74ALS139 was designed to be used primarily for memory decoding. Nevertheless, the 74ALS139 can be used in other applications as we'll see later in this section.

The 74ALS139 contains two independent 2-line to 4-line decoders in a single package. Each 2-line to 4-line decoder has 2 inputs and 4 outputs. In addition to the 2 inputs and to the 4 output pins, each 2-line to 4-line decoder has an enable input pin used to enable or disable one decoder or the other. A detailed description of the operation of the decoder pins appears later in this section

Figure 4.24 represents the 74ALS139. Pin number 1 or 1 $\overline{G}$ represents the enable input for the first decoder; pin numbers 2 and 3 (or 1A and 1B respectively) represent the inputs to the first decoder; pin numbers 4, 5, 6, and 7 (or 1Y0, 1Y1, 1Y2, and 1Y3 respectively) represent the outputs from the first decoder; pin number 8 (or GND) is the ground input; pin numbers 9, 10, 11, and 12 (or 2Y3, 2Y2, 2Y1, and 2Y0 respectively) represent the outputs from the second decoder; pin numbers 13 and 14 (or 2B and 2A respectively) represent the inputs to the second decoder; pin number 15 (or 2 $\overline{G}$) represents the enable input for the second decoder; and finally pin number 16 (or Vcc) is the 5 volt input.

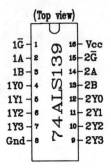

Figure 4.24. 74ALS139

Figure 4.25 represents the logic diagram for the 74ALS139 decoder. Note that the 74ALS139 contains two 2-line to 4-line decoders. Note the bubbles at inputs 1 $\overline{G}$ and 2 $\overline{G}$ and the bubbles at all the outputs. The bubbles indicate that the signal is low true.

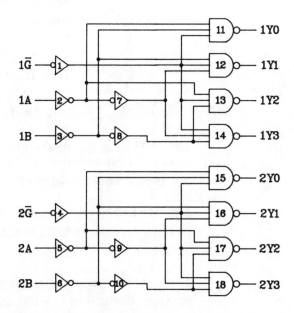

Figure 4.25. Logic diagram for the 74ALS139 decoder

The function table for the 74ALSf139 decoder appears in figure 4.26.

First Decoder

Inputs			Outputs			
Enable	Select					
$1\overline{G}$	1B	1A	1YO	1Y1	2Y2	2Y3
H	X	X	H	H	H	H
L	L	L	L	H	H	H
L	L	H	H	L	H	H
L	H	L	H	H	L	H
L	H	H	H	H	H	L

Second Decoder

Inputs			Outputs			
Enable	Select					
$2\overline{G}$	2B	2A	2YO	2Y1	2Y2	2Y3
H	X	X	H	H	H	H
L	L	L	L	H	H	H
L	L	H	H	L	H	H
L	H	L	H	H	L	H
L	H	H	H	H	H	L

X=Don't care   H=High   L=Low

Figure 4.26. Function table for the 74ALS139 decoder

Decoder Operation

Note from the logic diagram (figure 4.25) and from the function table (figure 4.26) the first and second decoders are exactly the same. Therefore the study of one decoder (first or second) could be extended to the second. We will discuss the operation of the first decoder.

$1\overline{G}$=High

The first decoder is disabled. From the logic diagram of the first decoder (figure 4.25), we see that applying a high logic signal (or logic 1) to input $1\overline{G}$ cause the first decoder to be disabled. Note that the high logic signal applied to $1\overline{G}$ is first inverted by gate 1 and becomes a low logic signal. Then the low logic signal emerging from gate 1 goes to gates 11, 12, 13, and 14 (*see* figure 4.25). Gates 11, 12, 13, and 14 are NAND gates. (Recall that in order for a NAND gate's output to be low, all of the NAND gate's inputs must be high.) In the case ($1\overline{G}$=high), at least 1 input of gates 11, 12, 13, and 14 is low. Therefore the output of gates 11, 12, 13, and 14 is automatically high regardless of which logic level (high or low) is applied to the other inputs of gates 11, 12, 13, and 14 (*see* figure 4.25). The "X's" in the first line (under 1A and 1B) in figure 4.26 mean that regardless of the logic signal (high or low) applied to inputs 1A and 1B, the outputs (1Y0, 1Y1, 1Y2, and 1Y3) are all high if the $1\overline{G}$ input signal is high. This is consistent with our earlier discussion of the logic diagram.

1 $\overline{G}$=Low and 1A=1B=Low

Output 1Y0 is activated. By applying a low logic signal to input 1$\overline{G}$, the decoder is enabled and inputs 1A and 1B will determine which output (1Y0, 1Y1, 1Y2, or 1Y4) will be activated. When a low logic signal is applied to both 1A and 1B, only output 1Y0 is activated. The output signal from gate 11 is low, and the output signals from gates 12, 13, and 14 are high. (*See* figure 4.25).

1 $\overline{G}$=Low and 1A=High 1B=Low

Output 1Y1 is activated. When a low logic signal is applied to 1B and a high logic signal is applied to 1A (with 1 $\overline{G}$=L), output 1Y1 is activated. In figure 4.25 we see that when a low logic signal enters gate 3 (or 1B) and a high logic signal enters gate 2 (or 1A), the output logic signals from gates 11, 13, and 14 (respectively 1Y0, 1Y2, and 1Y4) are high (or logic 1), and the output logic signal from gate 12 (or 1Y1) is low (or logic 0). Therefore output 1Y1 is activated. (Recall that the output logic is low true.)

1 $\overline{G}$=Low and 1A=Low 1B=High

Output 1Y2 is activated. If a low logic signal is applied to 1A and a high logic level is applied to 1B, output 1Y2 is activated. Using figure 4.25 we see that a high logic signal in 1B (gate 3) results in a low signal at gates 11 and 12. Therefore 1Y0 and 1Y1 are disabled. Also the low signal applied to 1A (gate 2) is inverted by gate 2 (becomes high logic) and inverted again by gate 7 (becomes low logic). The output signal (low) of gate 7 disables gates 14 and 12. Gate 12 is also disabled by the output of gate 3. Therefore 1Y3 (or gate 14) is disabled. On the other hand, gate 13 (or 1Y2) is activated (all its inputs are high), and the output of gate 13 is low.

1 $\overline{G}$=Low and 1A=1B=High

Output 1Y3 is activated. When a high logic signal (or logic 1) is applied to 1A and 1B (gates 2 and 3 respectively), gates 11, 12, and 13 are disabled. Note from figure 4.25 that the output from gate 2 is low (if input 1A is high). This low output signal from gate 2 disables gates 11 and 13. Also when the input to gate 3 is high, its output is low. The output of gate 3 (low) disables gate 12. Recall that one low input signal (or more) to a NAND gate is sufficient to force the output (of the NAND gate) to a high state (disabled). On the other hand, gate 14's inputs are all high when a high logic signal is applied to both 1A and 1B. Therefore the output of gate 14 (or 1Y3) is low, and output 1Y3 is said to be activated.

Description of the Circuit to Be Tested

Figure 4.27 illustrates the circuit we'll be testing. Note that only one of the two 2-line to 4-line decoders that the 74ALS139 contains will be tested.

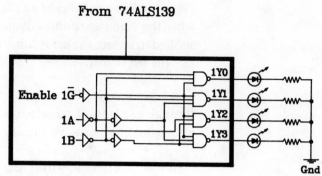

Figure 4.27. Circuit to be tested

Knowing that the outputs of the 2-line to 4-line decoder are low true, the leds of figure 4.27 will operate as follows:

Outputs	Output signals	Leds
Activated	Low	Off
Not activated	High	On

Figure 4.28. Low true logic led operation

From 4.28 we see that the leds will be "on" when the outputs are not activated, and the leds will be "off" when the outputs are activated. (Recall that the decoder outputs are low true, consequently a low output signal means that the output is activated, and a high output signal means that the output is not activated.) It is more practical to have the leds "on" when the outputs are activated, and to have the leds "off" when the outputs are not activated. Therefore we need to invert the output signals (from the 2-line to 4-line decoder) using NOT gates as follows:

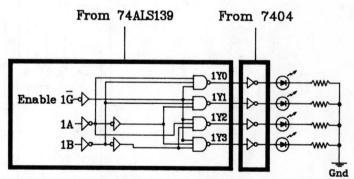

Figure 4.29. Final circuit to be tested in the experiment

Now that the output logic of the 2-line to 4-line decoder is high true, the leds of figure 4.29 operate as follows:

Outputs	Output signal	NOT gate	Leds
Activated	Low	High	on
Not activated	High	Low	off

Now, the leds are "on" when the outputs are activated, and the leds are "off" when the outputs are not activated. Recall that when a high signal (or 5 volts) is applied to an led, the led is "on." When a low signal (or 0 volts) is applied to an led, the led is "off."

The resistors (R1, R2, R3, and R4) are used to limit the flow of the current through the leds. If no resistors were used, the current flowing through the led's would be excessive causing damage.

Circuit to Be Tested

Figure 4.30 offers a wiring diagram of the circuit we'll be testing in our experiment. Recall that only one of the two 2-line to 4-line decoders will be tested. Therefore switches A, B, and G were wired to inputs 1A (pin #2), 1B (pin #3), and $1\overline{G}$ (pin #1) respectively. (*See* figure 4.30.)

When an led (L0, L1, L2, or L3) is on, we know that its corresponding output (1Y0, 1Y1, 1Y2, or 1Y3 respectively) was activated.

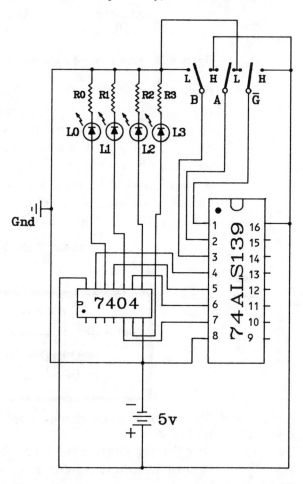

Figure 4.30. Diagram of the circuit to be tested

Testing the 74ALS139 Dual 2-line to 4-line Decoder

1 Wire the circuit shown in figure 4.30. (Refer to earlier experiments for wiring procedures). Remember to connect the battery or power supply last. Once the power source is connected, one led may be on; ignore this and go to step 2.

2 Now, move the $\overline{G}$ switch to the low position to enable the decoder. Switch $\overline{G}$ should be kept in the low position throughout the first part of the experiment (steps 2 through 6). At this time ignore the leds.

3 Move switches A and B to the low position and note which led's are on.

4 Move switch A to the high position (keep switch B in the low position), and record the result.

5 Move switch A to the low position and switch B to the high position. Record the result.

6 Move switches A and B to the high position. Record the result.

7 Move switch $\overline{G}$ to the high position (keep switch B in the high position), and record the result.

Results

Step 3
Only led 0 is on. Therefore the 1Y0 output was activated.

Step 4
Only led 1 is on. Therefore the 1Y1 output was activated.

Step 5
Only led 2 is on. Therefore the 1Y2 output was activated.

Step 6
Only led 3 is on. Therefore the 1Y3 output was activated.

Step 7
When input switch $\overline{G}$ is in the high position, the decoder is disabled. Therefore the position of switches A and B would not matter (because the decoder is disabled). All the leds are off.

Practical Applications of 2-line to 4-line Decoders

Figure 4.31 shows how a 74ALS139 can be interfaced to memory chips. Recall that the 74ALS139 decoder is a dual 2-line to 4-line decoder. For correct operation, only one of the 8 memory banks should be enabled at one time. Therefore only one of the decoder's outputs should be activated at any time. In figure 4.31, we disabled the second decoder by applying a high signal to pin $2\overline{G}$. We enabled the first decoder by applying a low signal to pin $1\overline{G}$ (*see* figure). Now we can only access MBI through MB4. Both the outputs from the decoder and the memory enable pins are low true. In other words when a particular output is activated, its corresponding memory bank is also activated. For example if we want to access MB2, output 1Y1 must be activated. If we want to access MB2, using figure 4.26 we see that a high signal should be applied to the A input (1A in our example), and a low signal should be applied to B input (1B in our example).

Figure 4.32 shows the connections between the memory chips and the decoder that concern our purpose; other connections and other memory pins are omitted. A summary of the operation of the circuit of figure 4.32 is given in figure 4.31. Refer to figure 4.26 for details on how the *output enabled* column was derived given the *inputs*.

1A	1B	1$\overline{G}$	2A	2B	2$\overline{G}$	Output enabled	Memory bank enabled
X	X	H	X	X	H	none	none
L	L	L	X	X	H	1Y0	MB1
H	L	L	X	X	H	1Y1	MB2
L	H	L	X	X	H	1Y2	MB3
H	H	L	X	X	H	1Y3	MB4
X	X	H	L	L	L	2Y0	MB5
X	X	H	H	L	L	2Y1	MB6
X	X	H	L	H	L	2Y2	MB7
X	X	H	H	H	L	2Y3	MB8

Figure 4.31. Summary of circuit operation

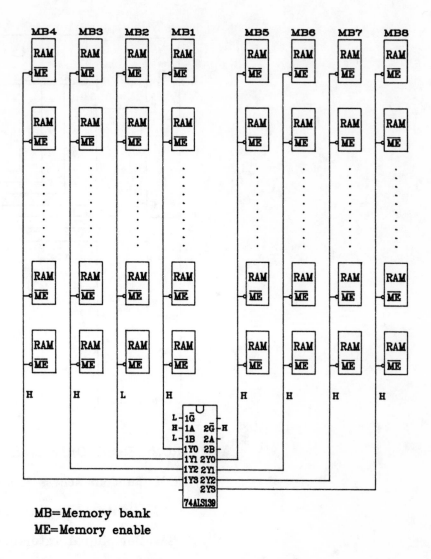

Figure 4.32. Interfacing the 74ALS139 with memory chips

Another example of a practical use for a 2-line to 4-line decoder is in a CPU board. Figure 4.33 shows how a CPU can communicate correctly with the PIA, ROM, RAM, and ACIA. Note that the CPU can communicate with only one component at any given time. Therefore it is necessary to disable all the components that are not accessed by the CPU.

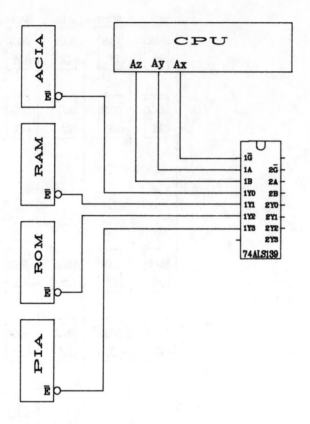

ROM: Read Only Memory

RAM: Random Access Memory

CPU: Central Process Unit

PIA: Peripheral Interface Adapter

ACIA: Asynchronous Communication Interface Adapter

Figure 4.33. Decoding components in a CPU board

The decoder (74ALS139) in figure 4.33 is used by the CPU to enable or disable the different components of the board. When the CPU needs to address a particular component, the CPU will generate an appropriate code (from its address lines represented in figure 4.33 as Ax, Ay, and Az). The code generated by the CPU (from Ax, Ay, and Az) will instruct the decoder (74ALS139) to enable a particular component.

Example

If for examaple the CPU wants to address the PIA, the CPU must generate 011 (Ax, Ay, and Az respectively).

Explanation

Note that the Ax pin in the CPU (figure 4.33) is wired to the 1 $\overline{G}$ pin of the 74ALS139 decoder. Therefore when the Ax pin of the CPU goes low, the 74ALS139 decoder is enabled (refer to figure 4.26.) Now let's look at the Ay and Az pins of the CPU (figure 4.33). We see that the Ay and Az pins are wired to the 1A and 1B pins of the 74ALS139 decoder. Therefore when pins Ay and Az of the CPU go high, the 1A and 1B pins of the 74ALS139 decoder also go high. Therefore output 1Y3 is activated (*see* figure 4.26). From figure 4.33 we see that output 1Y3 is wired to the enable pin of the PIA (E). Therefore when output 1Y3 is activated (low), the PIA is enabled and can be addressed by the CPU.

Conclusion

The codes that must be generated by the CPU to address a particular component are given in figure 4.34.

CPU Decoder

Ax (1 $\overline{G}$)	Ay 1A	Az 1B	Output line activated	Chip enabled
1	X	X	none	none
0	0	0	1Y0	ACIA
0	1	0	1Y1	RAM
0	0	1	1Y2	ROM
0	1	1	1Y3	PIA

Figure 4.34. CPU's codes to address components

Problems

Problem 4.1

Define a decoder.

Problem 4.2

Design a combinational circuit (decoder) which outputs binary 1 when it detects the presence of more than two binary 0's in its three binary outputs.

Problem 4.3

Which output is activated when the binary number 1000 is entered into the 7442 decoder? Explain.

Problem 4.4

What decimal number is displayed when the binary number 0100 is entered into a 7447 BCD to seven segment decoder? (Assume that the 7447 is wired to a 7 segment led display.)

Problem 4.5
Define an encoder.

Problem 4.6
State the differences between an encoder and a decoder.

Problem 4.7

Explain the principles of *priority* in the 74147 10-line-to-4-line encoder.

Problem 4.8

What output code is generated when a low signal is applied to the 3 and 8 inputs of a 74147 encoder? Explain.

Problem 4.9

What logic gates (OR, AND, or NOT) are to be used to interface the 74147 encoder to the 7447 BCD to 7 segment decoder?

Problem 4.10

Design a combinational circuit that would display the decimal values 0 to 9. (Use MSI decoders and encoders.)

5

Multiplexers/Demultiplexers

Introduction

Switches are often required in electronic applications. Although mechanical switches are used in electronic applications, these become useless when high speed switching is required. *Multiplexers* and *demultiplexers* are electronic switches that operate in the same manner as mechanical switches. They operate much faster however.

Multiplexers and demultiplexers can be designed using analog or digital components. Analog multiplexers/demultiplexers may be designed with bipolar or MOSFET (Metal Oxide Semiconductor Field Effect Transistor) switches. Digital multiplexers/demultiplexers are designed with logic gates. In this book, only digital switches (multiplexers/demultiplexer) will be covered.

Multiplexers, which are also referred to as *data selectors*, are used in many applications such as converting data from parallel to serial, generating binary words, and Boolean functions. Operation of these multiplexers will be discussed later in this chapter.

Multiplexers can be interpreted as one way rotary switches. A multiplexer selects one input signal (from many input signals) and directs it to a single output. Figure 5.1 shows a mechanical and an electronic (multiplexer) switch.

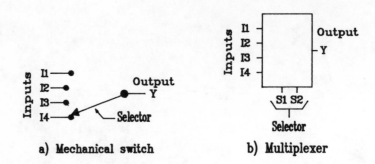

Figure 5.1. Mechanical and digital switches

The input lines to both the mechanical and the electronic switches are called I1, I2, I3, and I4. The output from both switches (mechanical and multiplexer) is called Y.

With the mechanical switch, the selector is used to select the input to be directed to the output. If the selector is in the I2 position, input signal I2 is routed to the output Y (*see* figure 5.1). If the selector is in position I4, input signal I4 is directed to the output (Y),..etc.

The multiplexer depicted in figure 5.1 has 2 selector lines (S1 and S2) which have the same function as the mechanical selector. By applying a logic level signal (0 or 1) on the S1 and S2 lines, one input signal (I1, I2, I3, or I4) is selected and directed to the output Y.

Note that a multiplexer must have as many select input lines as necessary to select and direct any one input line from all of its input lines. An 8 input multiplexer requires 3 select lines to be able to transfer (to the output) any of one of the 8 input lines. A 16 input multiplexer requires 4 select lines,...etc.

A demultiplexer, or *data distributor* is another switch that operates like a one way rotary switch. A demultiplexer operates in a manner exactly opposite that of a multiplexer. While a multiplexer has multiple input lines and a single output, a demultiplexer has a single input and multiple output lines. Figure 5.2 shows a one way rotary switch that corresponds to one particular demultiplexer.

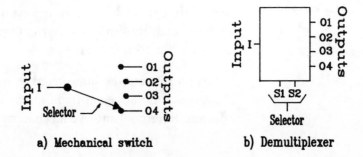

Figure 5.2. Mechanical and digital switches

The selector in the one way rotary switch selects the output which the input (I) is to be directed to. If the selector is in the O1 position, the input I is directed to **output line O1** (*see* figure 5.2). If the selector is in position O2, input 12 is directed to output line O2,..etc.

In the demultiplexer, S1 and S2 have the same role as the selector in the one way rotary switch. S1 and S2 are used to select the output line to be used to direct the input signals. The logic operation for a demultiplexer will be discussed later.

A demultiplexer must have as many input selectors as necessary to access all its outputs. (if the demultiplexer has 8 output lines, 3 selector lines are required; with 16 output lines a demultiplexer would require 4 selector lines,..etc.)

Multiplexers

A multiplexer (also called a data selector) is an electronic circuit used to select one input signal (from many) and direct that input to a single output.

A popular multiplexer is shown in figure 5.3. The 74ALS151 is an *8-line to 1-line multiplexer*. By 8-line to 1-line we mean that the multiplexer selects one input from eight available inputs and directs the selected input to a single output. The 8-lines (or inputs) of the 74ALS151 are known as *data sources*. These are represented by D0 through D7 (*see* figure 5.3). The single output of the 74ALS151 is represented by Y (*see* figure 5.3). As we'll see later, the W output is merely the complement of the Y output.

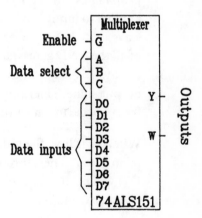

Figure 5.3. The 74ALS151 multiplexer

The A, B, and C inputs are called the *data select*. By applying an appropriate binary signal to A, B, and C, a particular data source (D0, D1, D2,..., or D7) will be selected. Once the data source is selected, the multiplexer directs the selected input to output Y. The operation of data select will be discussed in more detail later in this section.

The $\overline{G}$ input is called the enable input. It is used to enable or disable the 74ALS151 multiplexer as we'll discuss later.

The logic diagram of the 74ALS151 is presented in figure 5.4.

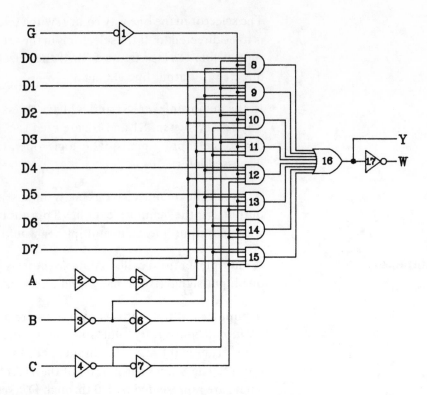

Figure 5.4. Logic diagram of the 74ALS151 multiplexer

If a high logic signal (or logic 1) is applied to the $\overline{G}$ output, the signal first propagates through gate 1 (*see* figure 5.4). Gate 1 complements the signal applied to input $\overline{G}$. Therefore the signal becomes low logic (or logic 0) when it leaves gate 1. Notice from figure 5.4 that the output of gate 1 is tied to one input of all the AND gates (gates 8 through 15). Therefore when $\overline{G}$ is high, the output of gate 1 is low. All the AND gates (gates 8 through 15) are inhibited by the output signal emerging from gate 1, thus preventing any of the input signals from appearing at output Y.

When $\overline{G}$ is low, the multiplexer is enabled by the high output signal emerging from gate 1. Therefore an input signal from one data input (D0, D1,..., D6, or D7) will be directed to output Y. (The operation of selecting one data input is given next in this section.)

The data select inputs A, B, and C (*see* figure 5.4) are used to select which data input (D0, D1,..., D6, or D7) is to be routed to output Y. Now, suppose that the multiplexer is enabled ($\overline{G}$=low), and the binary number 0 0 0 is applied to the data select (respectively C, B, and A). Input D0 will be selected and directed to output Y. The reason why D0 is selected and directed to output Y when the binary number 000 is entered into the data select follows:

Data Select (CBA)=000
When binary 000 is applied to the data select (CBA), with $\overline{G}$=low, the logic levels of the AND gates in the 74ALS151 accept the values shown in figure 10.5.

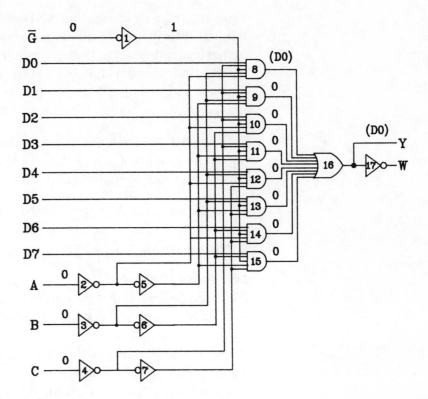

Figure 5.5. Gate's logic levels when selecting D0

Figure 5.6 summarizes the AND gate's logic inputs and outputs when the binary number 000 is applied to the data select CBA.

Gate #	Input(s)	Output logic
1	0	1
2	0	1
3	0	1
4	0	1
5	1	0
6	1	0
7	1	0
8	(D0)1111	(D0)
9	(D1)1011	0
10	(D2)1110	0
11	(D3)1010	0
12	(D4)1011	0
13	(D5)1010	0
14	(D6)1010	0
15	(D7)0010	0
16	(D0)0000000	(D0)=Y
17	(D0)	($\overline{D0}$)=W

(D_n) = logic value of D_n can be logic 0 or 1
D_n = D0, D1,..., or D7

Figure 5.6. AND gate's logic levels when input D0 is selected

From figure 5.6 we can see that AND gates 9 through 15 each contain at least one 0 at their inputs. Therefore the output for gates 9 through 15 is 0. The output from gate 8 will be the logic value of D0; either logic 0 or 1. If D0=0, the inputs to gate 8 are 01111; the output from gate 8 is then 0 (0=D0). If D0=1, the inputs to gate 8 are 11111; the output from gate 8 is then 1 (1=D0). The logic level of D0 will be output through gate 8.

Gate 16 (OR gate) will also transmit the logic level of D0 because its other inputs (the outputs of gates 9 through 15) are low. Therefore if D0 is low, the output from gate 16 is low; and if D0 is high, the output of gate 16 is high.

In conclusion, when the binary number 000 is entered into the data select, data input D0 is selected. Once D0 is selected, D0's logic level (logic 0 or 1) is directed to output Y.

Data Select (CBA)=001

Notice (in figure 5.4) that the output of gate 7 is tied to one input of gates 15, 14, 13, and 12. The output of gate 6 is tied to one input of gates 15, 14, 11, and 10. Finally notice that the output of gate 2 is tied to gates 14, 12, 10, and 8. When the binary number 001 (CBA respectively) is applied to the data select, logic 0 appears at the output of gates 7, 6, and 2. Therefore AND gates 8 and 10 through 15 are inhibited. The only AND gate not inhibited is gate 9. Four of the five inputs to AND gate 9 are high (outputs from gates 1, 3, 4, and 5). Thus the logic signal (low or high) of the fifth input to AND gate 9 (which is D1) will appear at output Y.

If D1=0, the output of gate 9 will be 0; the output of gate 16 will also be 0 (Y=0). If D1=1, the output of gate 9 will be 1, and the output of gate 16 will also be 1 (Y=1).

Data Select (CBA)=010

When 010 is applied to data select input lines CBA respectively, the following gates are inhibited:

1 Input C is low (or 0). Therefore the output of gate 7 is low and gates 15, 14, 13, and 12 are inhibited.

2 Input B is high (or 1). Therefore the output of gate 3 is low and gates 13, 12, 9, and 8 are inhibited.

3 Input A is low (or 0). Therefore the output of gate 5 is low and gates 15, 13, 11, and 9 are inhibited.

The only AND gate not inhibited is gate 10. Note that four of the five inputs of gate 10 are high. The output of gates 1, 2, 4, and 6 are inputs to gate 10. The fifth input to gate 10 is D2. Therefore the logic signal in D2 (low or high) appears at output Y when the binary number 010 is applied to the data select input lines CBA.

Data Select (CBA)=011

When 011 is applied to data select lines CBA respectively, the result is:

1 Input C is low (or 0). Therefore the output of gate 7 is low and gates 15, 14, 13, and 12 are inhibited.

2 Input B is high (or 1). Therefore the output of gate 3 is low and gates 13, 12, 9, and 8 are inhibited.

3 Input A is high (or 1). Therefore the output of gate 2 is low and gates 14, 12, 10, and 8 are inhibited.

The only AND gate not inhibited is gate 11. Four of the five inputs of gate 11 are high. The outputs of gates 1, 4, 5, and 6 comprise four of the five inputs to gate 11. The fifth input to gate 11 is D3. Therefore the logic signal in D3 will appear at output Y when the binary number 011 is applied to the data select input lines CBA.

We can continue testing the multiplexer data select input line by applying the binary numbers 100, 101, 110, and 111 to the data select. For each of these binary numbers one data input (D4, D5, D6, or D7 respectively) will appear at output Y. We can summarize the operation of the 74ALS151 multiplexer in the following function table (figure 5.7):

Inputs				Outputs	
Enable	Data Select				
G	C	B	A	Y	W
H	X	X	X	L	H
L	L	L	L	D0	D0
L	L	L	H	D1	D1
L	L	H	L	D2	D2
L	L	H	H	D3	D3
L	H	L	L	D4	D4
L	H	L	H	D5	D5
L	H	H	L	D6	D6
L	H	H	H	D7	D7

L=Low   H=High   X=Don't care

D0, D1...D7 = logic level of the respective D input

Figure 5.7. Function table for the 74ALS151 multiplexer

Example

Obtain the logic pulse that will appear at output Y when binary 011, 000, 111, and 010 are entered (in that order) into the data selector (CBA) of the multiplexer shown in figure 5.8.

(cont. on page 196)

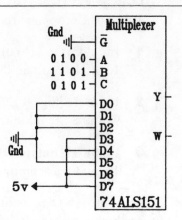

Figure 5.8. Multiplexer when binary 011, 000, 111, and 010 are
entered into the data selector

Solution

Recall from earlier experiments that 5 volts represents logic level 1, and
ground represents logic level 0. From figure 5.8 we notice that the $\overline{G}$
enable pin is connected to the ground. Therefore the 74ALS151
multiplexer is enabled (*see* figure 5.7).

1 Enter 011 into the data select.

When binary 011 is entered into the data select CBA, the data input
D3 is selected (*see* figure 5.7). Therefore the logic level of D3 is
transmitted to output Y. The data input D3 is wired to 5 volts, and 5
volts is also binary 1. Thus, when 011 enters the data select (CBA), 5
volts is transmitted to output Y. Output Y is said to be high logic, or
output Y is said to be equal to binary 1.

2 Enter 000 into the data select.

When binary 000 is entered into the data select CBA, D0 is selected
(*see* figure 5.7). Therefore the logic level of D0 is transmitted to
output Y. D0 is wired to the ground. Thus 0 volts are transmitted to
output Y. Output Y is said to be low logic, or output Y is said to be
equal to binary 0.

3 Enter 111 into the data select.

When binary 111 is entered into the data select CBA, D7 is selected
(*see* figure 5.7). Therefore the signal in D7 is transmitted to output
Y. From figure 10.8, we see that D7 is wired to 5 volts. Therefore 5
volts are transmitted to output Y. Output Y is said to be high logic, or
output Y is said to be equal to binary 1.

4 Enter 010 into the data select.

Data input D2 is selected when binary 010 is entered into the data
select CBA (*see* figure 5.7). The signal in data input D2 is trans-

(cont. on page 197)

> mitted to output Y. From figure 5.8 we see that D2 is wired to ground. Therefore 0 volts are transmitted to output Y. Output Y is said to be low logic, or output Y is said to be equal to binary 0.

The logic pulse that appears at output Y when binary 011, 000, 111, and 010 are entered (in that order) into the data select CBA is shown in figure 5.9. Notice that output W is the complement of output Y.

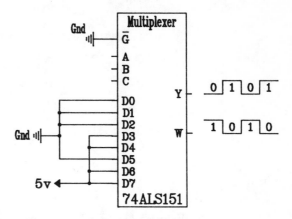

Figure 5.9. Output pulse

**Practical
Applications
with the
74ALS151
Multiplexer**

**Parallel to Serial
Data Conversion**

As we mentioned in the introduction, a multiplexer has other applications beside data selection. Multiplexers are used for other purposes such as parallel to serial data conversion, binary word generation, and finally Boolean function implementation.

Suppose that a parallel binary word "b7 b6 b5 b4 b3 b2 b1 b0" is applied to the data input lines D7 D6 D5 D4 D3 D2 D1 D0 of the 74ALS151 multiplexer (*see* figure 5.10). If the input select states of CBA are stepped from 000 through 111, inputs D0 through D7 will be enabled in sequence producing a serial binary word as shown in figure 5.10.

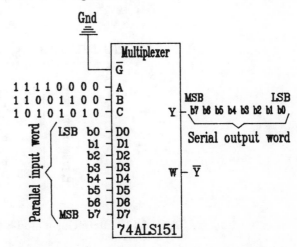

Figure 5.10. Parallel to serial data conversion with the 74ALS151

Note that b0 represents the logic level in D0, b1 represents the logic level in D1,...etc.

When 000 is entered into the data sheet, D0 is selected (*see* figure 5.7), and b0 is directed to Y. When 001 is entered into the data select, D1 is selected, and b1 is directed to Y,...etc. Figure 5.11 represents the logic level that output Y takes when the data select inputs C, B, and A are stepped from 000 through 111. (Refer to figure 5.7.)

Data select			Input selected	Output	
C	B	A		Y	
0	0	0	D0	b0	(LSB)
0	0	1	D1	b1	
0	1	0	D2	b2	
0	1	1	D3	b3	
1	0	0	D4	b4	
1	0	1	D5	b5	
1	1	0	D6	b6	
1	1	1	D7	b7	(MSB)

Figure 5.11. Serial output word sequence

In figure 5.10, we've converted the parallel input word "b7 b6 b5 b4 b3 b2 b1 b1" into an equivalent serial output word, demonstrating that the 74ALS151 is indeed capable of converting a parallel word to a serial word.

A modulo 8 binary counter can be used to generate the binary sequence 000 through 111 (refer to chapter 2: Counters). If a mod-8 binary counter is connected to the data select of the 74ALS151 as shown in figure 5.12, there is no need to enter binary 000 through 111 into the multiplexer's data select.

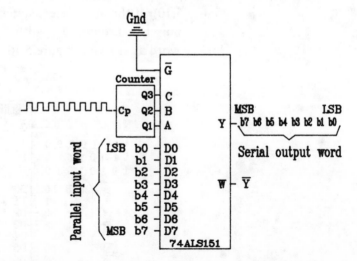

Figure 5.12. Interfacing a mod-8 binary counter to the 74ALS151

From figure 5.12, we notice that the counter's Q3 output is wired to the multiplexer's C input. Q2 (counter's output) is wired to the B (multiplexer's input). Finally Q1 (counter's output) is wired to A (multiplexer's input). We learned in chapter 2 that a mod-8 counter count sequence is: 000, 001, 010, 011, 100, 101, 110, and 111. Thus when the counter is counting, data select inputs C, B, and A are stepped from 000 through 111, and the parallel to serial conversion take place.

Example

What binary word appears at output Y when pulses are applied to the CP input (counter) as shown in figure 5.13. Explain!

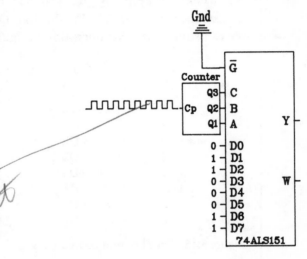

Counter will provide output for all bit sequences

Figure 5.13. Data conversion with a multiplexer

Solution

When pulses are entered into the CP input of the counter in figure 5.13 the counter's outputs Q3, Q2, and Q1 will take the following values:

Pulse #	Outputs		
	Q3	Q2	Q1
1	0	0	0
2	0	0	1
3	0	1	0
4	0	1	1
5	1	0	0
6	1	0	1
7	1	1	0
8	1	1	1

Figure 5.14. Counter's outputs

(cont. on page 200)

Note from figure 5.13 that the outputs of the counters Q3, Q2, and Q1 are wired to the C, B, and A (respectively) data select lines of the 74ALS151 multiplexer. Therefore when pulse #1 is applied to the CP input of the counter (*see* figure 5.14), the outputs of the counter Q3, Q2, and Q1 are respectively 000. Therefore data input D0 is selected and directed to Y (*see* figure 5.7). The logic level of D0 is 0 (or low) (*see* figure 5.13). Therefore a logic 1 is directed to output Y. When the second pulse is applied to the CP input of the counter in figure 5.14, the counter outputs are 001 (Q3 Q2 Q1 respectively). Therefore data input D1 is selected and directed to output Y. The logic level of D1 is 1. Therefore a logic 1 is directed to output Y.

As we continue applying pulses to the CP input of the counter, the logic levels appearing at inputs D2 through D7 will be directed to output Y. We can summarize this operation in figure 5.15.

Pulse #	Output activated	Logic of output Y
1	D0	0
2	D1	1
3	D2	1
4	D3	0
5	D4	0
6	D5	0
7	D6	1
8	D7	1

Figure 5.15. Serial output of figure 5.9

Figure 5.16 represents the result of the above operation. Note that by applying pulses to the CP input of the counter (figure 5.13), the binary number 11000110 was converted from parallel to serial (*see* figure 5.16).

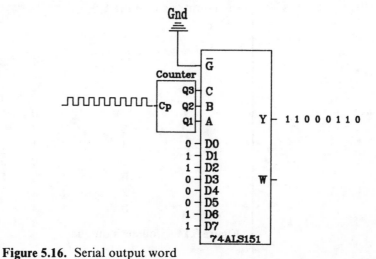

Figure 5.16. Serial output word

**The 74ALS151
as a Binary
Word Generator**

Some special digital circuits require the generation of a fixed single serial binary word. A multiplexer (74ALS151 for example) can be used to feed a fixed serial binary word to such a special circuit.

Suppose a special circuit must be fed the serial binary value 10010011. The 74ALS151 can be wired into the circuit so that the serial binary word 10010011 is generated.

**Design
Procedure**

We have to connect each data input (D0...D7) to 5 volts or to ground so that the fixed serial binary number 10010011 can be generated by the multiplexer when input select states C B A are stepped from 000 through 111. We know that 5 volts is equivalent to binary 1 and 0 volts (or ground) is equivalent to binary 0.

The first binary bit to reach the special circuit is the rightmost binary bit (or LSB) of the binary word 10010011. The rightmost binary bit is 1. Therefore data input D0 should be connected to 5 volts, so that when binary 000 is entered into data select CBA, bit 1 will be directed to Y, and then fed to the special circuit. (*See* figure 5.18).

The second bit to enter the special circuit is also 1. Therefore data input D1 must be connected to 5 volts, so that when the binary 001 is entered into the data select CBA, bit 1 will be directed to Y, and then fed into the special circuit.

The third bit to enter the special circuit must be 0. Thus data input D2 should be connected to 0 volts (or ground), so that when binary 010 is entered into the data select CBA, bit 0 will be directed to Y, and then fed into the special circuit...etc. We can summarize this operation in figure 5.17.

Data input	Connection	Data select C	B	A	Output Y	
D0	5 volts	0	0	0	1	(LSB)
D1	5 volts	0	0	1	1	
D2	Ground	0	1	0	0	
D3	Ground	0	1	1	0	
D4	5 volts	1	0	0	1	
D5	Ground	1	0	1	0	
D6	Ground	1	1	0	0	
D7	5 volts	1	1	1	1	(MSB)

Figure 5.17. Summary of binary word 10010011 generation

Figure 5.18 represents a 74ALS151 multiplexer wired so that the single fixed serial binary word 10010011 is generated each time that the data select states are stepped from 000 through 111.

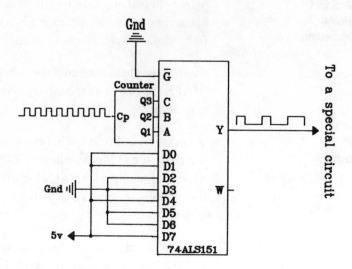

Figure 5.18. A fixed word generator

The 74ALS151 as a Boolean Function Generator

The 74ALS151 multiplexer can be wired so that it generates a Boolean function in the sum-of-products form when its data select states (CBA) are stepped from 000 through 111.

We see that the products $\overline{A}\,\overline{B}\,\overline{C}$ through ABC are developed by gates 8 through 15, as shown in figure 5.19.

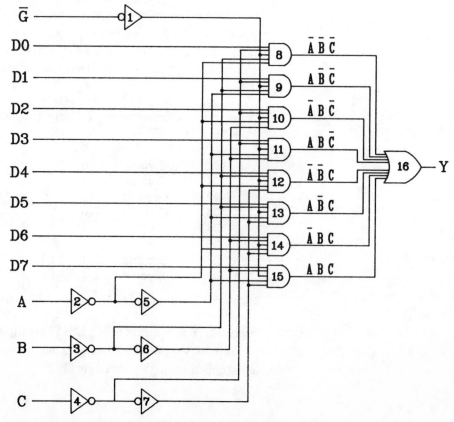

Figure 5.19. Output from gates 8 through 15

To select a product, a binary 1 (or 5 volts) should be applied to the appropriate data input. **For example if the product ABC (*see* figure 5.19) is to appear in** the Boolean function, data input D7 must be connected to 5 volts.

If a product is to be erased from the Boolean function, binary 0 (or ground) must be applied to the appropriate data input. If for example the product ABC should not be part of the Boolean function, D7 should be connected to ground.

In the following example, we'll wire a 74ALS151 multiplexer so that the Boolean function $Y = AB\overline{C} + \overline{A}\,\overline{B}C + \overline{A}BC + ABC$ is generated when the data select states (CBA) are stepped from 000 through 111.

Gates 8, 9, 10, and 13 should be inhibited. We don't want the products $AB\overline{C}$. $\overline{A}\,\overline{B}C$, $\overline{A}BC$, and ABC to be part of the Boolean functions. Therefore data inputs D0, D1, D2 and D5 must be wired to 0 volts (or ground) to disable gates 8, 9, 10, and 13. (*See* figure 5.19).

Gates 11, 12, 14, and 15 must be enabled so that the 74ALS151 multiplexer generates their corresponding products (*see* figure 5.19) when the data select states (CBA) are stepped from 000 through 111. Therefore data inputs D3, D4, D6, and D7 must be connected to 5 volts.

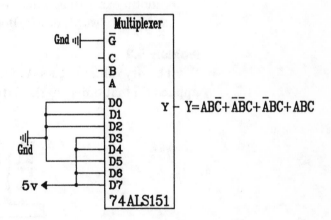

Figure 5.20. Multiplexer wired as a Boolean function generator

Figure 5.21 shows the values that output Y takes when the data select states (CBA) are stepped from 000 through 111.

C	B	A	Y
0	0	0	0
0	0	1	0
0	1	0	0
0	1	1	$AB\overline{C}$
1	0	0	$\overline{A}\,\overline{B}C$
1	0	1	0
1	1	0	$\overline{A}BC$
1	1	1	ABC

Figure 5.21. Output products

Therefore when the data select states (CBA) are stepped from 000 through 111, the following Boolean function is generated:

$$Y = 0 + 0 + 0 + AB\overline{C} + \overline{A}\overline{B}C + 0 + \overline{A}BC + ABC$$
$$= AB\overline{C} + \overline{A}\overline{B}C + \overline{A}BC + ABC$$

Practice Problems

Problem 5.1

State the 3 functions of a multiplexer

Solution

A multiplexer is a combinational circuit that can be used as:

1 An electronic switch.

A multiplexer is used to select one input (from many), and directs the input's logic signal (binary 0 or 1) to output Y.

2 A data convertor.

Because of its many inputs and its single output, a multiplexer can be used to convert parallel data into serial data.

3 Word and Boolean function generator.

A multiplexer can be wired so that it (the multiplexer) generates any single fixed binary word, or any Boolean function.

Problem 5.2

What binary value appears at output Y when binary 011, 110, 100, and 010 are applied (in that order) to the data selector as shown in figure 5.22.

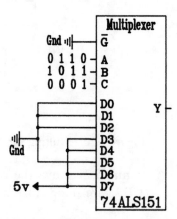

Figure 5.22. Multiplexer wired as a binary word generator

Solution

Binary 011 is the first binary value to enter the data select (CBA). Therefore data input D3 is selected (*see* figure 5.7). Data input D3 is connected to 5 volts (figure 5.22). Thus binary 1 (or 5 volts) is transmitted to output Y.

When binary 110 is entered into the data select (CBA), data input D6 is selected (*see* figure 5.7). Data select D6 is connected to 5 volts (*see* figure 5.22). Thus binary 1 (or 5 volts) appears at output Y.

When binary 100 is entered into the data select (CBA), data input D4 is selected (*see* figure 5.7). Therefore the logic signal in D4 is transmitted to output Y. From figure 5.22 we see that D4 is connected to 5 volts (or binary 1). Thus binary 1 (or 5 volts) appears at output Y.

Finally when binary 010 is entered into data select (CBA), D2 is selected (*see* figure 5.7) D2 is connected to 0 volts (or ground). Therefore binary 0 (or 0 volts) appears at output Y. Figure 5.23 represents the output pulse.

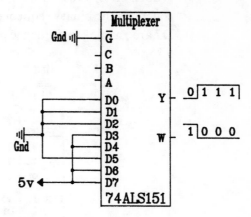

Figure 5.23. Output pulse

Problem 5.3

Design a multiplexer that will convert the parallel input word 10100011 into serial form.

Solution

The 74ALS151 can be used to convert parallel data into serial data. Figure 5.24 shows how a multiplexer can be used to convert the parallel input word 10100011 into a serial form.

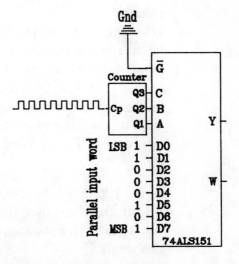

Figure 5.24. Data conversion with the 74ALS151

Note that the binary number to be converted was applied to the data inputs D0 through D7. The least significant bit (or LSB) was applied to data input D0, and the most significant bit (or MSB) was applied to D7 (*see* figure 5.24). The LSB was applied to D0 because D0 is the first data input to be enabled. (The first binary value to enter the data select CBA is 000, and binary 000 enables D0.) The MSB was applied to D7 because D7 is the last data input to be enabled. (The last binary that enters data select CBA is 111, and binary 111 enables D7.) Now when pulses are applied to the counter, the counter generates the binary sequence 000 through 111. Thus data inputs D0 through D7 are enabled in sequence producing the serial output word shown in figure 5.25.

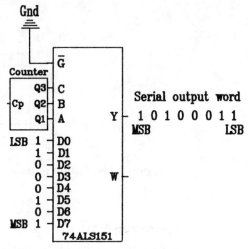

Figure 5.25. Serial output word

Problem 5.4

What serial output word would be output if the counter in figure 5.24 was a *down* counter?

Solution

When pulses are entered into a down counter, the binary sequence generated by the counter (Q3 Q2 Q1) decrements from 111 to 000. Therefore the first binary to enter the data select (CBA) is 111 (and not 000 as with an up binary counter). Thus the first data input to be enabled is D7 (and not D0 as with an up counter). The second binary to be generated by the down counter is 110 (Q3 Q2 Q1). Therefore when 110 enters into the data select (CBA), D6 is enabled. (*See* figure 5.27)...etc. The data input's enabling sequence proceeds from D7 to D0.

Figure 5.26 shows the serial output word. Note (from figure 5.26) that when a down counter is used to select the data inputs, the parallel input word is inverted when it becomes a serial word (LSB takes the place of the MSB, and vice versa).

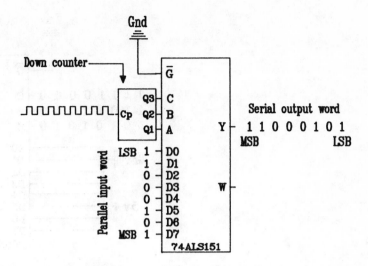

Figure 5.26. Serial output word with a binary down counter

If the LSB of the binary number was applied to data input D7 and if the MSB of the binary number was applied to data input D0, the serial output word would not have been inverted.

Problem 5.5
Wire a 74ALS151 multiplexer so that the fixed serial binary word 11110000 is generated.

Solution
The 74ALS151 multiplexer can be used as a serial word generator. Data inputs D0 through D7 should be wired to 5 volts or to 0 volts depending on the word to be generated. (Recall that high=5 volts=binary 1; and low=ground=binary 0.)

The rightmost 4 bits of the binary word 11110000 are 0000. Therefore data inputs D0, D1, D2, and D3 must be wired to ground (or 0 volts). Now when binary 000 through 011 enter the data select CBA, D0 through D3 are enabled, and four binary 0's appear at output Y. (Refer to figure 5.7 for the data input's enabling codes.)

To output the four binary 1's, D4, D5, D6, and D7 must be wired to 5 volts. Now, when binary 100 through 111 are entered into the data select CBA, D4 through D7 are enabled and 4 binary 1's appear at output Y (refer to figure 5.7).

Figure 5.27 represents a multiplexer wired so that the binary word 11110000 is generated when the data select states are stepped from 000 through 111.

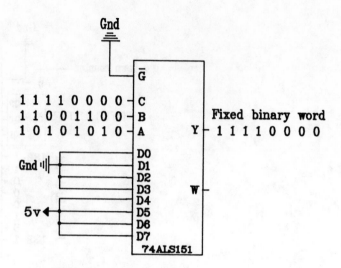

Figure 5.27. Serial binary word generation

Problem 5.6

Wire a 74ALS151 multiplexer so that the Boolean function f= $\overline{A}\,B\overline{C}$ + AB$\overline{C}$ + A$\overline{B}$C is generated when binary 000 through 111 are entered into the data select CBA.

Solution

From figure 5.19, we see that gate 10 develops $\overline{A}\,B\overline{C}$, gate 11 develops AB$\overline{C}$, and finally gate 13 develops A$\overline{B}$C. Therefore in order to develop the Boolean function f= $\overline{A}\,B\overline{C}$ + AB$\overline{C}$ + A$\overline{B}$C gates 10, 11, and 13 must be enabled. Data inputs D2, D3, and D5 are inputs to gates 10, 11, and 13. Thus, in order to enable gates 10, 11, and 13; data inputs D2, D3 and D5 must be connected to 5 volts.

Again from figure 5.19, we see that gates 8, 9, 12, 14, and 15 develop $\overline{A}\,\overline{B}\,\overline{C}$, A$\overline{B}\,\overline{C}$, $\overline{A}\,\overline{B}$C, $\overline{A}$BC, and ABC respectively. $\overline{A}\,\overline{B}\,\overline{C}$, A$\overline{B}\,\overline{C}$, $\overline{A}\,\overline{B}$C, $\overline{A}$BC, and ABC must not be included in the output Boolean function. Thus gates 8, 9, 12, 14, and 15 must be inhibited (or disabled). To inhibit gates 8, 9, 12, 14, and 15, data inputs D0, D1, D4, D6, and D7 which are inputs to gates 8, 9, 12, 14, and 15 respectively, must be connected to ground.

Figure 5.28 represents the 74ALS151 wired so that the Boolean function f=$\overline{A}$B$\overline{C}$ + AB$\overline{C}$ + A$\overline{B}$C is generated when the data select states CBA are stepped from 000 through 111.

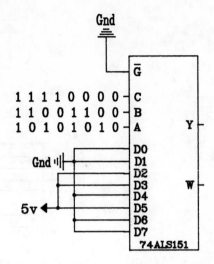

Figure 5.28. Wiring the 74ALS151 to generate a Boolean function

Data select			Gate accessed	Gates output
C	B	A		
0	0	0	8	0
0	0	1	9	0
0	1	0	10	$\overline{A}B\overline{C}$
0	1	1	11	$AB\overline{C}$
1	0	0	12	0
1	0	1	13	$A\overline{B}C$
1	1	0	14	0
1	1	1	15	0

Figure 5.29. Summary of operation

The OR gate (gate 16) of figure 5.19 adds the outputs of gates 8 through 15. The result of the addition is visible at gate 16's output (or Y). The outputs of gates 8 through 15 are represented in figure 5.29. When the data select states CBA are stepped from 000 through 111 we arrive at:

$$Y = f = 0 + 0 + \overline{A}B\overline{C} + AB\overline{C} + 0 + A\overline{B}C + 0 + 0$$

$$f = \overline{A}B\overline{C} + AB\overline{C} + A\overline{B}C$$

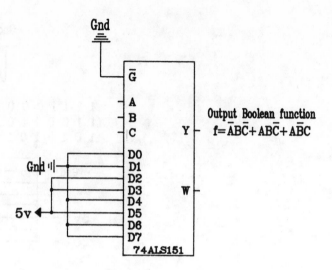

Figure 5.30. Output Boolean function

Demultiplexers

A demutiplexer (also called data distributor) is a combinational circuit that is the reverse of a multiplexer. While a multiplexer selects one input (from many) and directs the logic signal of the selected input to its single output; a demultiplexer selects one *output* (from many), and directs the logic signal of its single input to the selected output. In other words a multiplexer has multiple inputs and a single output, while a demultiplexer has a single input and multiple outputs.

Figure 5.31 represents the logic diagram of a demultiplexer. The demultiplexer depicted in figure 5.31 is a 1-line to 4-line demultiplexer. By 1-line to 4-line demultiplexer we mean that the demultiplexer selects 1 output (from the 4 output lines O1, O2, O3, or O4), and transmits the signal on its single input D to the selected output.

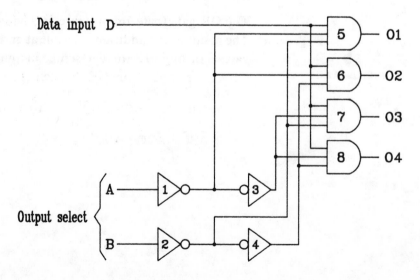

Figure 5.31. 1-line to 4-line demultiplexer

By applying a binary number to the output select, A and B, one output (O1, O2, O3, or O4) is selected. In other words one AND gate (5, 6, 7, or 8) is enabled when a binary number is applied to the output select (an explanation follows).

When a low signal (or binary 0) is applied to both A and B, gate 5 is enabled and gates 6, 7, and 8 are inhibited.

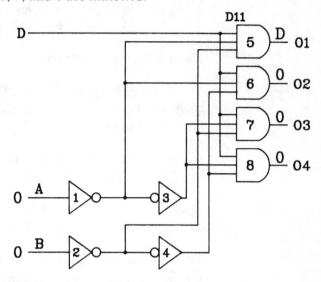

Figure 5.32. Enabling gate 5

Explanation

Tracing the signals as shown in figure 5.32 from A B to gate 5, we see that the inputs to gate 5 are D11 (where 'D' represents the signal applied to the data input). Thus the output signal of gate 5 is the same as the signal applied to the data input D. If the signal applied to D is binary 0, the output signal of gate 5 is also binary 0; if the signal applied to D is binary 1, the output signal of gate 5 is also binary 1. Because the logic signal (binary 0 or 1) in D is going to be transmitted through gate 5, gate 5 is said to be enabled. When binary 00 enters A B, output O1 is selected to carry the signal applied to the data input D.

Tracing the signals from A B to gate 6 as shown in figure 5.32, we see that gate 6's inputs are D10 (when binary 00 enters A B). Therefore the output signal from gate 6 is 0 no matter what signal is applied to the data input D (one binary 0 is sufficient to disable an AND gate). Gate 6 is inhibited (or disabled) when binary 00 is applied to output select lines A and B.

If we trace the signals from A B to gate 7 as shown in figure 5.32, we see that gate 7's inputs are D01. The binary 0 at gate 7's inputs forces gate 7's output (O3) to logic 0. Therefore gate 7 is inhibited when binary 0 0 enters A B.

Finally, let's trace the signals from A B to gate 8. When binary 00 is applied to A B, gate 8's inputs are D00 (*see* figure 5.32). Therefore gate 8's output is 0 no matter what logic signal is applied to the data input D. Gate 8 is said to be inhibited (or disabled) when binary 00 is applied to À B.

Figure 5.33 summarizes the output select operation when binary 00 through 11 are applied to A B (you can trace them using figure 5.32).

Output select		Gate 5		Gate 6		Gate 7		Gate 8	
A	B	In	Out	In	Out	In	Out	In	Out
0	0	D11	D	D10	0	D01	0	D00	0
0	1	D10	0	D11	D	D00	0	D01	0
1	0	D01	0	D00	0	D11	D	D10	0
1	1	D00	0	D01	0	D10	0	D11	D

Figure 5.33. Output selection

From figure 5.33, we see that when binary 00 is applied to AB, gate 5 is enabled. Therefore output O1 is said to be selected. When O1 is selected, the logic signal in the data input D is transmitted to O1.

When binary 01 is applied to A B, gate 6 is enabled. Thus output O2 carries the logic signal applied to the data input D, and output O2 is said to be selected.

With binary 10 applied to A B, gate 7 is enabled. The logic signal applied to the data input D is transmitted to output O3. Output O3 is said to be selected.

Finally when binary 11 is applied to A B, gate 8 is enabled. Therefore the signal applied to the data input D is transmitted to output O4. Output O4 is said to be selected.

An enable/disable input line could be implemented to the logic circuit of figure 5.31. Figure 5.34 illustrates how an enable input could control the demultiplexer depicted in figure 5.31. If a high signal is applied to the G input, a low signal would be transmitted from gate 9. Gate 9's output is an input to all the AND gates (5 through 8). Therefore all AND gates are inhibited, and the demultiplexer is said to be disabled (or shut off).

If a low signal is applied to the $\overline{G}$ input, a high signal will be transmitted from 9 to all AND gates (*see* figure 5.34). Consequently output select A B determines which AND gate is to be enabled. Now the demultiplexer is said to be enabled and its operation is represented in figure 5.33.

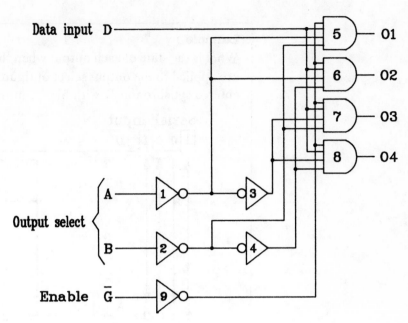

Figure 5.34. 1-line to 4-line demultiplexer with an enable input

Figure 5.34 represents a multiplexer that operates exactly as the demultiplexer depicted in figure 5.31 when a low signal is applied to the $\overline{G}$ input; but the demultiplexer can be shutoff (or disabled) when a high logic signal is applied to the $\overline{G}$ input.

Example

What binary number should be applied to the output select (A and B) in order to connect the data input D to output O3 (use figure 5.31)?

Solution

From figure 5.31 we see that output O3 is the output of gate 7. In order to enable gate 7, binary 10 should be applied to the output select (AB). (*See* figure 5.33.) Once gate 7 is enabled, the logic signal applied to the data input D would be transmitted to output O3. In other words, data input D is connected to output O3 when binary 10 is applied to the output select.

Example

What is the state of each output when the binary numbers 00 through 11 are applied to the output select of figure 5.35?(Suppose that the pulse is entered synchronously with binary numbers 00 through 11.)

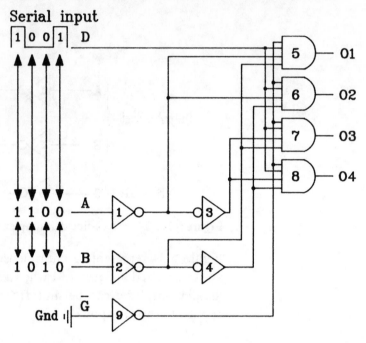

Figure 5.35. Demultiplexer operation

Solution

From figure 5.35 we see that the G input is tied to 0 volt (or ground). Therefore the demultiplexer is enabled (refer to figure 5.34).

From figure 5.34 we see that the demultiplexer input signals (AB and D) are synchronized as follows: when binary 00 and 11 are applied to the output select (AB), binary 1 is applied to the data input D. When 01 and 10 are applied to the output select (AB), binary 0 is applied to the data input D.

Order of entry	Output select		Data input
	A	B	D
First	0	0	1
Second	0	1	0
Third	1	0	0
Fourth	1	1	1

(cont. on page 215)

1 **Binary 00 enters the output select.**

When binary 00 enters the output select (A and B), gate 5 is enabled (refer to figure 5.33), and the signal applied to the data input D is transmitted to output O1. Therefore binary 1 (which is applied to 'D' when binary 00 enters AB) is transmitted from data input 'D' to output O1. Output O1=binary 1.

2 **Binary 01 enters the output select.**

Gate 6 is enabled when binary 01 is applied to the output select (refer to figure 5.33), and the signal applied to the data input 'D' is transmitted to output O2. Thus binary 0 (which is applied to 'D' when binary 10 enters AB) is transmitted from data input 'D' to output O2. Output O2=binary 0.

3 **Binary 10 enters the output select.**

When binary 10 enters the output select, gate 7 is enabled (refer to figure 5.33). Thus the signal applied to the data input 'D' is transmitted to output O3. Therefore binary 0 is transmitted from data input 'D' to output O3. Output O3=binary 0.

4 **Binary 11 enters the output select.**

When binary 11 enter the output select gate 8 is enabled (refer to figure 5.33). Thus the logic signal applied to the data input 'D' is transmitted to output O4. Therefore binary 1 is transmitted from data input 'D' to output O4. Output O4=binary 1.

In conclusion, we notice that the serial input data 1001 applied to the data input D was converted to parallel output data 1001 (*see* figure 5.36). Therefore a demultiplexer can be used as a data convertor to convert data from serial to parallel.

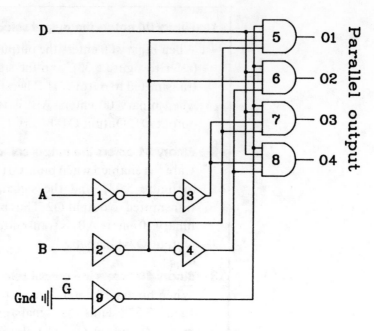

Figure 5.36. A demultiplexer as a data converter

Practice Problems

Problem 5.7

State the differences between a multiplexer and a demultiplexer.

Solution

Multiplexers and demultiplexers are combinational circuits that operate like mechanical switches. They are much faster however.

Multiplexers can be compared to one way rotary switches. Multiplexers select one input (from many inputs) and direct the signal applied to the selected input to the sole single output.

Demultiplexers can also be compared to one way rotary switches, but their operation is opposite that of multiplexers. Demultiplexers select one output (from many outputs), and direct the signal applied to the sole single input to the sole selected output.

Because multiplexers have multiple inputs and a single output, multiplexers are often used to convert parallel data (applied to the multiplexer's multiple inputs) into serial data (from the multiplexer's single output). On the contrary demultiplexers have a single input and multiple outputs. Demultiplexers are used to convert serial data (applied to the sole demultiplexer's input) to parallel data (from the demultiplexer's multiple outputs).

Problem 5.8

What binary value should be applied to the output select of the 1-line to 4-line demultiplexer depicted in figure 5.31 to transmit the signal applied to the data input D to output O2 ?

Solution

From the table of figure 5.33, we see that in order to enable gate 6 (whose output is O2), binary 01 must be applied to the output select A B. When binary 01 is applied to A B, gates 5, 7, and 8 are disabled. (Trace the signals from A B to gates 5, 7, and 8 using figure 5.31). At the same time gate 6 is enabled (trace the signal), and the signal applied to data input D is transmitted through gate 6 to arrive at output O2.

Problem 10.9

What parallel binary number is detected at outputs O1, O2, O3, and O4 when binary 10 (first), 01, 11, and 00 (last) are applied in that order to the output select A B as shown in figure 5.37.

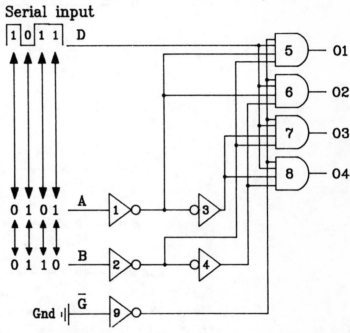

Figure 5.37. Demultiplexer

Solution

The first binary value that enters the output select A B is 1 0 (*see* figure 5.37). Using the table of figure 5.33, we see that when binary 1 0 enters A B, gate 7 is enabled. Now the signal applied to data input D is transmitted to output O3 (output O3 is the output of gate 7). The binary value applied to the data input D when binary 1 0 enters A B is 1 (*see* figure 5.37). Therefore binary 1 is transmitted to O3, and O3=1.

The second binary that enters A B is 0 1 (*see* figure 5.37). From the table depicted in figure 5.33, we see that gate 6 is enabled. When gate 6 is enabled, the signal applied to the data input D is transmitted to the output of gate 6 (or O2). From figure 5.37, we see that binary 1 is the signal that enters D when binary 0 1 enters A B (*see* figure 5.37). Therefore binary 1 is transmitted to O2, and O2=1.

The third binary to enter A B is 1 1. Using figure 5.33, we see that gate 8 is enabled when binary 11 enters A B. When gate 8 is enabled, the signal applied to D is transmitted to output O4. Binary 0 is applied to the data input D when binary 1 1 enters A B (*see* figure 5.37). Thus binary 0 is transmitted to output O4, and O4=0.

Finally, the fourth binary thay enters A B is 0 0. When binary 0 0 enters A B, gate 5 is enabled (*see* figure 5.33). Therefore the signal applied to the data input D is transmitted to output O1. Binary 1 is applied to the data input D when binary 0 0 enters A B (*see* figure 5.37). Thus binary 1 is transmitted to output O1, and O1=1. Figure 5.38 represents the binary outputs.

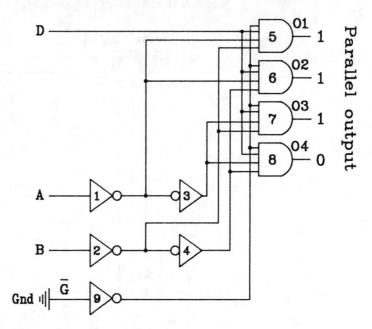

Figure 5.38. Demultiplexer's binary outputs

Problem 5.10

Repeat problem 5.9 with the binary values 00, 01, 10, and 11 (in that order) entering the demultiplexer as shown in figure 5.39.

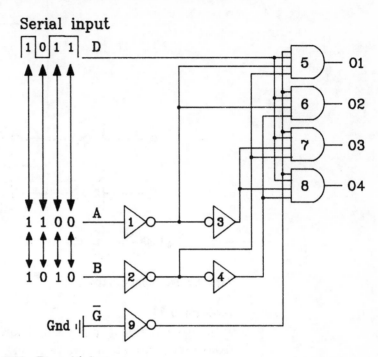

Figure 5.39. Demultiplexer

Solution

The first binary value to enter A B is 0 0 (*see* figure 5.39). When binary 0 0 enters A B, the binary value applied to D is transmitted to output O1 (*see* figure 5.33). Therefore binary 1 (which is the first binary applied to the data input D) is transmitted to output O1.

The second binary value to enter A B is 0 1. When binary 0 1 enters A B, gate 6 is enabled, and the binary applied to D is transmitted to output O2. Therefore binary 1 enters D when 0 1 enters A B (*see* figure 5.39), and binary 1 is transmitted to output O2.

The third binary value to enter A B is 1 0. From figure 5.33 we see that gate 7 is enabled when binary 1 0 enters A B. Thus the binary applied to the data input D when binary 1 0 enters A B is transmitted to output O3. From figure 5.39 we conclude that binary 0 is applied to the data input D when 1 0 enters A B. Therefore binary 0 is transmitted to output O3.

Finally, when binary 1 1 enters the output select A B, gate 8 is enabled (*see* figure 5.33). The signal applied to the data input D is transmitted to output O4 when gate 8 is enabled. Thus binary 1 which is the binary value applied to D when binary 11 enters A B (*see* figure 5.39) is (binary 1) transmitted to output O4. Figure 5.40 represents the output binary logic.

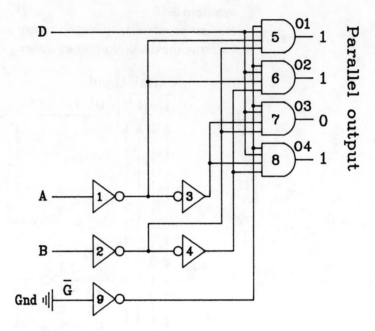

Figure 5.40. Binary outputs

Problem 5.11

What combinational circuit would you use to convert the binary number 1100 from a serial form to a parallel form?

Solution

Demultiplexers are widely used when a serial binary number is to be converted to parallel. Because the binary number 1100 is a 4 bit binary number, a 1-line to 4-line demultiplexer is needed. The 1-line to 4-line demultiplexer depicted in **figure 5.34 can be used for this application.**

We have to apply the serial number to be converted to the data input D of the demultiplexer. We must also apply binary 00 through 11 to the output select A B of the demultiplexer. Figure 5.41 shows the serial binary number applied to data input D, and binary 00 through 11 applied to the output select A B.

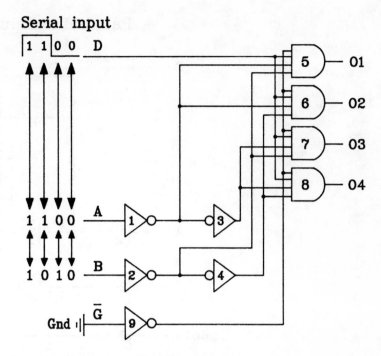

Figure 5.41. Application of a serial binary number

Now let's enter the binary values 00 through 11 to the output select A B, while the serial binary number 1100 is entered into the data input D as shown in figure 5.41.

The first binary value to enter A B is 0 0. Therefore binary 0 is transmitted to output O1 (*see* figure 5.33), and O1=0.

The second binary value to enter A B is 0 1. Thus binary 0 is transmitted to output O2 (*see* figure 5.33), and O2=0.

The third binary value to enter A B is 1 0. Thus binary 1 is transmitted to output O3 (*see* figure 5.33), and O3=1.

The last binary value to enter A B is 1 1. Therefore binary 1 is transmitted to output O4 (*see* figure 5.33), and O4=1.

The serial binary number 1100 has been converted to a parallel form (*see* figure 5.42). Note that O4 represents the MSB (or the most significant bit) of the serial binary number, and O1 represents the LSB (or the least significant bit) of the serial binary number.

Parallel output

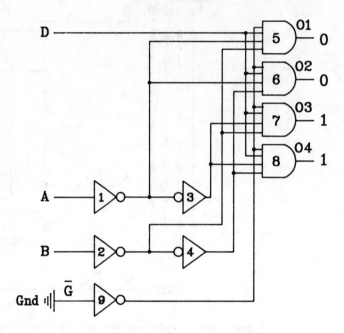

Figure 5.42. Parallel form of the serial number

Problems

Problem 5.1
Define a multiplexer.

Problem 5.2
Describe three applications of a multiplexer.

Problem 5.3
Define a demultiplexer.

Problem 5.4

Discuss the differences between a multiplexer and a demultiplexer.

Problem 5.5

Obtain the output pulse-train that emerges from output Y if binary 111, 000, 101 and 011 are entered (in that order) into the data selector (CBA) of a 74ALS151 multiplexer.

Problem 5.6

Using a 74ALS151 multiplexer design a circuit that will convert the parallel binary word 10010110 into a serial form.

Problem 5.7

Using a 74ALS151 multiplexer design a circuit that will generate the fixed serial binary number 01110001.

Problem 5.8

Using a 74ALS151 multiplexer design a circuit that will generate the following Boolean function:

$$f = (\overline{A}\,\overline{B}C) + (\overline{A}BC) + (A\overline{B}\,\overline{C}) + (A\overline{B}C) + (ABC)$$

Problem 5.9

Design a circuit that will convert the serial binary number 1001 into a parallel form.

Problem 5.10

What parallel binary number is detected at the outputs 01, 02, 03, and 04 if the binary numbers 01, 10, 11, and 00 are entered (in that order) into the output select AB of a 1-line to 4-line demultiplexer. Suppose that the serial binary number 1011 is sensed at the D input of the demultiplexer.

Index